THE DRIVING MACHINE

ALSO BY
WITOLD RYBCZYNSKI

Paper Heroes
Taming the Tiger
Home
The Most Beautiful House in the World
Waiting for the Weekend
Looking Around
A Place for Art
City Life
A Clearing in the Distance
One Good Turn
The Look of Architecture
The Perfect House
Vizcaya (with Laurie Olin)
Last Harvest
My Two Polish Grandfathers
Makeshift Metropolis
The Biography of a Building
How Architecture Works
Mysteries of the Mall
Now I Sit Me Down
Charleston Fancy
The Story of Architecture

The DRIVING MACHINE

A DESIGN HISTORY OF THE CAR

WITOLD RYBCZYNSKI

WITH DRAWINGS BY THE AUTHOR

W. W. NORTON & COMPANY
Independent Publishers Since 1923

Printed in the United States of America
First Edition

For information about special discounts for bulk purchases, please contact
W. W. Norton Special Sales at specialsales@wwnorton.com or 800-233-4830

Manufacturing by Lakeside Book Company
Book design by Barbara M. Bachman
Production manager: Anna Oler

ISBN 978-1-324-07528-8

W. W. Norton & Company, Inc.
500 Fifth Avenue, New York, N.Y. 10110
www.wwnorton.com

W. W. Norton & Company Ltd.
15 Carlisle Street, London W1D 3BS

10 9 8 7 6 5 4 3 2 1

To the memory of

Bing Thom (1940–2016)

Money may not buy happiness,

but I'd rather cry in a Jaguar than on a bus.

—FRANÇOISE SAGAN

CONTENTS

PREFACE

STANLEY DONEN'S 1967 FILM, *TWO FOR THE ROAD,* USES A series of cars to trace the rise of a couple's fortune. A hitchhiking Albert Finney is given a lift in a Volkswagen bus and meets Audrey Hepburn. Married, they tool around southern France in an old MG TD, the sports car self-destructs, and they graduate to a red Triumph Herald convertible, ending up in a splendid white Mercedes 230SL roadster. The film made an impression on me because Finney plays an architect—like me—and I too have a soft spot for cars.

Car design and architectural design have similarities; both are concerned with balancing the various and sometimes conflicting demands of function, construction, and aesthetics. Yet there are significant differences. An architect has to decide how to deal with what Frank Gehry once called the burden of history; whether to honor the past or ignore it. Car design is not like that. Once the link to the horse-drawn carriage was broken, which happened early, there was no preordained model—a car could be whatever you wanted it to be, boxy or streamlined, two-door or four-door, it could have four wheels or three. The automobile created its own mutating tradition. For decades the front grille, which masked the radiator, was an identifying feature; think of the Rolls-Royce temple, the BMW double kidney, or the Volvo slashed bar. Then, with the electric car, the grille abruptly disappeared.

Building design has many conventions: front doors open inward, corridors have standard dimensions, closets are two feet deep. Car design is less constrained: the engine can be in the front or the back, or in the middle; gear shifters can be on the floor, on the steering column, or sticking out of the dash; bumpers can be exposed or concealed; doors can swing open, slide, or even lift. Although some early cars had steering tillers, the wheel has persisted, but for how much longer? Some manufacturers have experimented with unconventional steering yokes, though so far without great success. Not that buyers aren't open to innovation—unusual cars such as the VW Beetle, Fiat Cinquecento, and Mini have been bestsellers—but buyers can also reject designs, as they did with the Chrysler Airflow, the Henry J, and the Ford Edsel. Car makers realized early that form did not simply follow function—automobile design was highly technical, but it also required an artistic sensibility. Enter the stylist—and, inevitably, fashion. There have been automotive baroque periods when chrome ruled, and minimalist eras when chrome disappeared. As in architecture, aesthetic sensibility and engineering demands have been tempered by changing tastes.

Cars, unlike buildings, have a relatively short life; in most states, a twenty-five-year-old vehicle is eligible for an antique license plate. Buildings, on the other hand, last for centuries. You can still walk into the lobby of the Chrysler Building on Lexington Avenue in Manhattan, but you will not find a Chrysler Airflow parked on your street. To see this car, you have to go to a car museum or a classic car show. Thus, old cars recede from memory. If they survive at all in the public's imagination, it is thanks to movies: the 1934 Ford V8 coupe in the gruesome final scene of *Bonnie and Clyde*; the stately 1940 Buick Phaeton Limited convertible that takes Rick and Ilsa to the airport in *Casablanca*; or the 1981 DeLorean in *Back to the Future*.

Cars are machines, but like buildings they are also cultural artifacts. The earliest automobiles often conveyed national traits: the down-to-earth Morris Minor; the no-nonsense Mercedes; the charming Fiat Topolino; the stylish Citroën DS. Since the 1980s, the best-selling vehicle in the United States has not been a car at all, but the Ford F-series pickup truck, a distinctive American vehicle that deserves its own design history. Like buildings, automobiles reflect cultural preoccupations. And fashions. Many cars of the thirties, such as the magnificent Czech Tatra and even the humble Volkswagen, were influenced by the Art Deco movement; the 1948 Jaguar XK120 inaugurated a new, glamorous era in car design; by the fifties, the influence of aeronautics appeared in cars as disparate as the Saab 92 and the gloriously finned 1957 Imperial Crown. Such great old cars deserve the same serious attention that we give to great old buildings—they are design achievements and an important part of our material past.

Over five decades, I've owned fifteen cars. Despite my frequent buying—I refuse to call it a compulsion—I wasn't one of those people who identified with their car; for me, it was more like a tool. But I did care about design. Using a well-designed tool, whether it was a Peugeot pepper mill, a Dupont fountain pen, or a Contax II camera, provided its own pleasures. A good tool is like a natural extension of the body, a kind of prosthesis. Somebody—usually many somebodies—has refined the tool until it does its job near perfectly. The driving machine and its controls are the result of more than a century of trial and error: the foot pedals, the gearshift lever, the instrument cluster, the rearview mirror. Everything is in its place. Using a well-designed tool, I feel connected to the past, whether I am swinging a hammer—a tool of ancient Roman origin—or sitting behind a steering wheel.

THE DRIVING MACHINE

CHAPTER
ONE

STARTING THE CAR

Well, sometimes Ledwinka looked over my shoulder and sometimes I looked over his.

—FERDINAND PORSCHE ON DESIGNING THE VOLKSWAGEN

MY FIRST CAR WAS A VOLKSWAGEN BEETLE. THE DESIGN was already thirty years old—seven years older than I—and production would continue for another thirty-six years, making it the world's longest-running automobile model. I bought mine in January 1967. After college, I'd worked and saved my money, and, as many young architects had done before me, I set out on a European grand tour. My VW was seven years old, bought for $300 in Hamburg, where the freighter on which I had booked passage from Quebec City berthed. The car carried me and a friend without a hitch from Paris to Valencia (where it was stolen, but that's another story). Nine hundred miles. I'd never driven a VW, but the controls were simplicity itself. The enameled metal dashboard was dominated by a large speedometer that included an odometer, three warning lights—blue for the high beams, red for the generator, and green for low oil pressure—and an indicator for the flashing turn signal lights, which by 1960 had replaced the original semaphore indicators. There was no temperature gauge because the engine was air-cooled. There was no gas gauge, either—when the engine stuttered, it meant the tank was empty, which required flipping a lever below the dash to access the reserve—about a gallon, good for roughly forty miles. The dashboard included two unidentified white plastic pull knobs: the one on the left controlled the headlights; on the right, the windshield wipers. I believe that there was also a choke. The turn signal was controlled by a stalk on the steering column, and the headlights were dimmed by depressing a floor switch. There was an ashtray, although no lighter. A knurled knob on the floor beside the stick shift controlled the heat, which came from an exchanger surrounding the exhaust pipe. The first time I stopped for gas, I looked in vain for the gas cap—I found it under the hood, which was actually a trunk, since the engine was in the rear.

I was reminded of my old VW the other day when my friend Jerry gave me a ride in his new Prius. Instead of a speedometer and traditional gauges, there was what Toyota calls a Multi-Information Display, an LCD touch screen swimming in icons, graphics, and numbers. The colorful array, which reminded me of a pinball machine, conveyed a range of technical information such as tire pressure and fuel consumption, as well as navigation and entertainment information and extraneous details such as time and date and whether a door was open. Somewhere, there was a number indicating the car's speed. The cost of adding digital information is minimal, and I had the impression that the designers had simply piled on the bells and whistles. Perhaps that's why Jerry's owner's manual was almost eight hundred pages (compared to ninety for a 1960 VW), twenty pages just on operating the lights and wipers. Jerry told me his dealer offered a course for new owners on how to manage the complicated display. No doubt one got used to the busy flat screen, but I would miss the elegant minimalism of that old VW.

The car I bought in Hamburg had distinctive oval German export license plates, and an international registration sticker marked with a *D*, for Deutschland. As I drove through Holland on my way to Paris, more than once when asking for directions I received dirty looks, especially from older persons for whom the wartime German occupation was a living memory. And, after all, my car's godfather was Adolf Hitler himself. Opening the 1933 Berlin Motor Show as the newly appointed reich chancellor, he had announced a national policy to motorize Germany, which despite having invented the automobile a half century earlier, lagged other European countries in car ownership. Hitler called on the auto industry to produce an affordable people's car, a *volkswagen*.

The automotive engineer who would realize Hitler's vision was not a German native. Ferdinand Porsche (1875–1951) was born in the small Bohemian market town of Maffersdorf in the Austro-Hungarian Empire; after the First World War, he would become a citizen of the newly created Czechoslovak Republic.* Young Porsche worked in his father's tinsmith shop and attended evening classes at the local polytechnic college. The boy was fascinated by electricity, and built his own generator, making the Porsche home the first in town to have electric light. At the age of eighteen, thanks to the recommendation of a local businessman, the precocious youngster was apprenticed to Béla Egger, a Viennese manufacturer of electrical equipment. Four years later, Egger began a collaborative project to develop an electric automobile with Jacob Lohner & Company, an established luxury coach builder. Porsche, who had risen in the company, was charged with the design of the motor and drivetrain.

At the time that Egger and Lohner started their project, it was far from clear which motive power was best suited to the automobile: steam, gas-fired internal combustion, or electricity. Many argued for steam, which had the advantage of being a tried and true technology—after all, James Watt's steam engine dated from 1776—and steam had been used to power tractors, omnibuses, and fire trucks since the mid-nineteenth century. Not only were steam engines safe and reliable, and mechanically straightforward, but steam could be produced by burning

* Porsche renounced his Czech citizenship in 1934, when he became a naturalized German citizen.

kerosene, which—unlike gasoline—was widely available. One of the pioneers was the Frenchman Amédée Bollée (1844–1917), whose steam-powered omnibus, called *L'Obéissante* (the Obedient One), could carry twelve passengers. In 1873, Bollée demonstrated its effectiveness by driving from his native Le Mans to Paris. It took him eighteen hours to make the 150-mile journey, with stops for meals, for replenishing the boiler, and not least for speeding tickets (the vehicle had a top speed of twenty-five miles per hour), although the fines were rescinded following Bollée's triumphant arrival in Paris.

The disadvantage of a steam-powered vehicle was the heavy boiler, which needed constant attention, and the external combustion steam engine, which was considerably less efficient than an internal combustion gasoline-powered engine. What is widely considered the world's first successful internal combustion–engine motorcar was built by Carl Benz (1844–1929) in 1885. The vehicle resembled a large tricycle: the padded bench was supported on a tubular-steel frame, and the three wire-spoked wheels had solid rubber tires. A single front wheel did the steering; the two large back wheels were powered by a chain drive connected to a small one-cylinder, four-stroke internal combustion engine of Benz's own design. Producing less than one horsepower, and stabilized by a horizontal flywheel, the engine was sufficient to power the six-hundred-pound vehicle at ten miles per hour. Steering was by means of a tiller, while a hand lever applied power or brakes—there was no throttle. A knob beneath the seat adjusted the carburetor, a device that mixed air and fuel prior to combustion. The Benz Patent-Motorwagen went into production the following year. Sales were slow, and to help generate publicity—and without Benz's knowledge—in August 1888 his wife, Bertha (1849–1944), whose dowry had financed the project, drove a prototype from Mannheim, where the family lived, to her native town of Pforzheim, sixty-five miles away. There were, of course, no gas stations, and she stopped at a pharmacy to purchase bottles of petroleum ether, a cleaning fluid, to use as

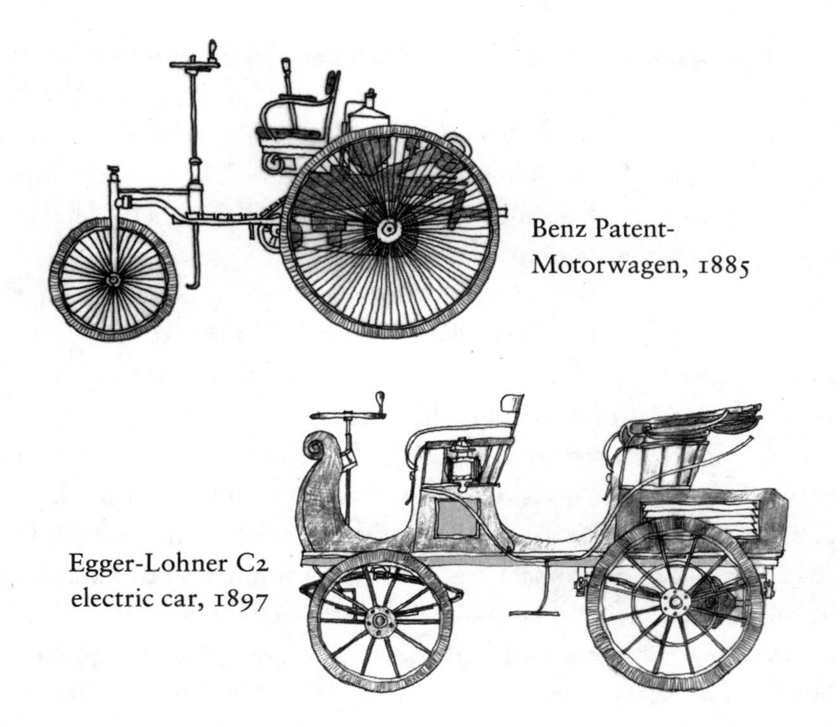

Benz Patent-Motorwagen, 1885

Egger-Lohner C2 electric car, 1897

fuel. Bertha was accompanied by her two teenage sons. The historic journey took twelve hours, and on uphill stretches the boys had to get out and push.

The internal combustion engine was not an overnight success. Cars powered by electric motors were quieter and simpler to build and operate since electric motors did not require gears. The Egger-Lohner electric car that the young Porsche worked on was a tall four-seater whose large wooden-spoked wheels, sprung body, and folding canvas top resembled a traditional phaeton coach—it would be a few more years before automobile bodies assumed a distinctive form. The large rear wheels were powered by a three-horsepower electric motor situated behind the axle, and the vehicle had a top speed of about twenty miles per hour. A box under the rear bench seat housed the batteries. Batteries were the electric car's weak spot. The Egger-Lohner's batteries provided a limited range—about fifty miles. Moreover, at a time when electrification was confined to cities, anyone taking a drive in the country risked getting stranded. And batteries were heavy; the Egger-Lohner's seventy-four-cell lead-acid batteries added up to almost a third of the vehicle's three-thousand-pound weight. That meant sluggish performance, especially when climbing hills, as well as heavy wear on the pneumatic tires.

The Egger-Lohner electric car was one of the first automobiles in the Austro-Hungarian Empire, but although four prototypes were built, the car did not go into production. Nevertheless, the Lohner company, recognizing Porsche's obvious talent, recruited him to be the head designer of its newly formed electric car division. Earlier, when he had been working at Egger, Porsche attended an evening class at Vienna's Technical University, and to ease his bicycle commute he built a small hub-mounted electric motor to drive the rear wheel. It was an impressive feat for a tinkering teenager. Porsche subsequently patented an electric hub motor suitable for automobiles. Driving the wheels directly with small motors reduced weight and did away with the need for a transmission, which increased efficiency. One of Lohner's early electric cars was a custom-built racing car for a wealthy Englishman that used four hub motors, one on each wheel. The ponderous vehicle, which included a massive battery weighing almost two tons, has been described as a battery box on wheels.

To reduce the battery size, and overcome the electric car's limited range, Porsche began to work on a car whose electric motor was backstopped by an internal combustion engine—in other words, a hybrid. Each of the two front wheels of the three-passenger vehicle was powered by a 2.7-horsepower electric hub motor that was connected to a forty-four-cell lead-acid battery, about half the size of the one in the Egger-Lohner car. The battery was housed in a box suspended beneath the steel chassis with shock-absorbing springs to safeguard the delicate lead plates. The battery was capable of powering the wheel-hub motors for about fifty miles, at which point the driver flipped a switch that started a pair of small one-cylinder water-cooled internal combustion engines located behind the driver's seat (the radiators were mounted on two sides of the front cowling). Each engine ran a generator that provided power to one of the hub motors as well as charging the battery. Once the battery was charged, the engines could be turned off. The rudimentary dashboard had two dials, a voltmeter and an ammeter. A lever varied the current flow, which controlled the speed; a steering wheel controlled the front wheels. The range of the hybrid vehicle, which had a top speed of about twenty miles per hour, was effectively unlimited, and Lohner named it the Semper Vivus—"Always Alive."

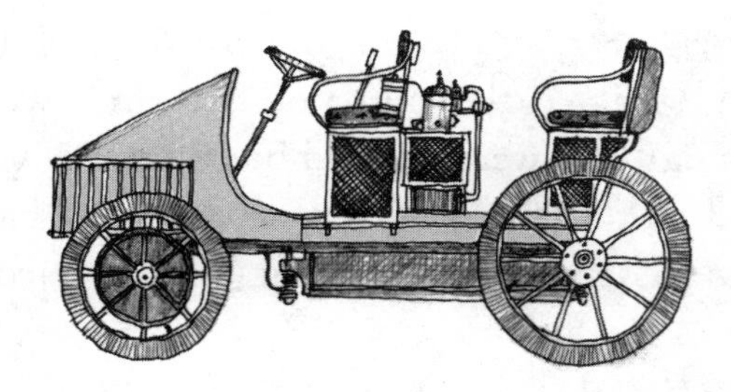

Lohner Semper Vivus hybrid, 1900

The ingenious Semper Vivus, which today would be called a "concept car," was displayed at the 1900 Paris Exposition, where it was a public sensation and won a gold medal. Lohner used the hybrid system on large vehicles such as fire trucks and buses, and also manufactured a passenger car version, although the expensive car sold primarily to wealthy enthusiasts. Porsche, a keen motor racer, built a successful hybrid racing car using hub motors on all four wheels.

Even as the Semper Vivus was making a splash, car design was evolving. In the early 1890s, the French firm Panhard et Levassor built several cars with front-mounted internal combustion engines driving the rear wheels. The cars used a clutch pedal and a sliding-mesh gearbox—that is, the gears on the main shaft were "shifted" to mesh with the appropriate gears on the countershaft.

This arrangement proved so successful that the so-called Système Panhard quickly became an industry standard. One of the leading car makers at that time was Daimler-Motoren-Gesellschaft (DMG), founded by the Stuttgart engineer Gottlieb Wilhelm Daimler (1834–1900). Daimler and his partner, Wilhelm Maybach (1846–1929), were pioneers of the internal combustion automobile. Daimler and René Panhard were friends, and DMG cars all used front-mounted engines with sliding-mesh transmissions. In 1901, the German company unveiled a car commissioned by Emil Jellinek (1853–1918), a DMG distributor, who suggested a sportier car than the current models. Often called the first modern car, the resulting product was the work of Maybach, who was DMG's chief engineer, and Paul Daimler (1869–1945), the founder's eldest son. Unlike earlier automobiles, which were influenced by carriage design and were tall and boxy, the DMG car was low and elongated, with a long hood that extended back smoothly to contain the passenger compartment. The car incorporated the

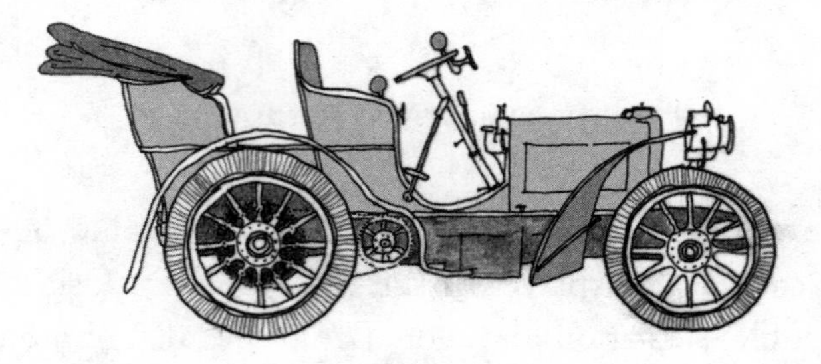

Daimler-Motoren-Gesellschaft Mercedes 35HP, 1901

first honeycomb radiator, designed by Maybach to improve cooling efficiency, and the prominent radiator grille was framed in chrome and emblazoned with the tag "Mercedes" (the name of Jellinek's daughter). The four-cylinder engine put out thirty-five horsepower, enabling the Mercedes 35HP to reach the previously unattainable speed of fifty miles per hour.

In 1906, Austro-Daimler, DMG's Austrian subsidiary, which had manufactured the gasoline engines for the Semper Vivus, recruited Ferdinand Porsche to be its head designer, succeeding

Paul Daimler, who had returned to Germany. Porsche stayed with the company almost two decades and eventually became managing director, overseeing the transformation of Austro-Daimler into an independent company. During this time, he continued to be drawn to racing. Motor racing had developed hand in hand with the automobile; races were popular with the public, and they enabled manufacturers to promote their products. The most prestigious race in Germany was the annual Prinz Heinrich Tour, a thousand-mile rally, held in six stages from Berlin to Bad Homburg, near Frankfurt. It was an endurance race, limited to four-seater production cars, and the rules required drivers to be accompanied by three passengers. In 1910, Austro-Daimler entered eight of its new 27/80 touring cars, large four-seaters with folding tops that had been developed specifically for the rally. The cars used the Système Panhard. The internal combustion engine, designed by Porsche, was a 5.7-liter four-cylinder that produced ninety-five horsepower and powered the car to an impressive eighty miles per hour. An

Austro-Daimler 27/80 Prinz Heinrich, 1910

innovation was the location of the camshaft, which converted the reciprocating motion of the pistons into rotation. Porsche moved the shaft from below the combustion chamber to above. The so-called overhead cam had fewer moving parts, produced more power, and yielded better fuel consumption. The speedy Austro-Daimler touring cars placed first, second, and third, with Porsche himself driving the first-place car.

The 27/80, renamed the Prinz Heinrich after its victory, had a distinctive one-piece body that entirely shrouded the passenger compartment and terminated in a bullet-shaped rear that contained the gas tank. The design was the work of Ernst Neumann-Neander (1871–1954), a German painter and graphic artist who owned an advertising agency that specialized in motorcars and motorcycles, which led him to work as a design consultant on cars and motorbikes, making him one of the first automobile stylists. The Prinz Heinrich's so-called torpedo body marked an important moment in automotive design. In the dozen years that had passed since the coach-like Egger-Lohner car, the automobile had evolved a distinctive form, an expression not of stately motion, but of speed.

The depressed post–World War I economy of the dissolved Austro-Hungarian Empire hit automobile manufacturing hard, and in 1923, Porsche left struggling Austro-Daimler and joined DMG in Stuttgart. The southwestern German city, which was the home of automotive pioneers such as Daimler, Maybach, and Benz, had become the center of the German automotive industry. Porsche became DMG's technical director, again succeeding Paul Daimler, who had left after a disagreement with the board. Five years later, DMG merged with Benz, which was the largest German car maker, to become Daimler-Benz. Its cars were named Mercedes-Benz, the Mercedes being DMG's most successful model. Ferdinand Porsche was a perfectionist and not always easy to work with, and a year after the merger, he had a falling-out with the new management. It is unclear whether he quit or was fired, but in any case he and his family returned to Vienna. Porsche briefly joined Steyr, an Austrian car maker, and in 1931, when the Great Depression took hold and the local economy collapsed, the peripatetic designer, now unemployed, returned to Stuttgart.

This time, Porsche did not go to work for a car company. At the relatively advanced age of fifty-six, he founded an independent consulting firm specializing in engine and automobile design. His staff included several of his Austrian and Czech engineer colleagues, as well as his son, Ferry. One of the firm's first automobile projects was the design of a budget car for Zündapp, a motorcycle company that was looking to expand into automobiles. The result was a small four-seater. Pioneering car designers such as Porsche were in uncharted waters, and as the Prinz Heinrich shows, they were free to innovate. While hardly a racing car, the Zündapp T12 had a top speed of fifty miles per hour, and to improve performance, the Porsche engineers gave the car a streamlined shape with a curved front and a sloping rear. The small engine, which was air-cooled and did not require a radiator, was located in the rear, which eliminated the need for a driveshaft. Three prototypes of the Zündapp were built, but when the development costs mounted, the company got cold feet and halted the project.

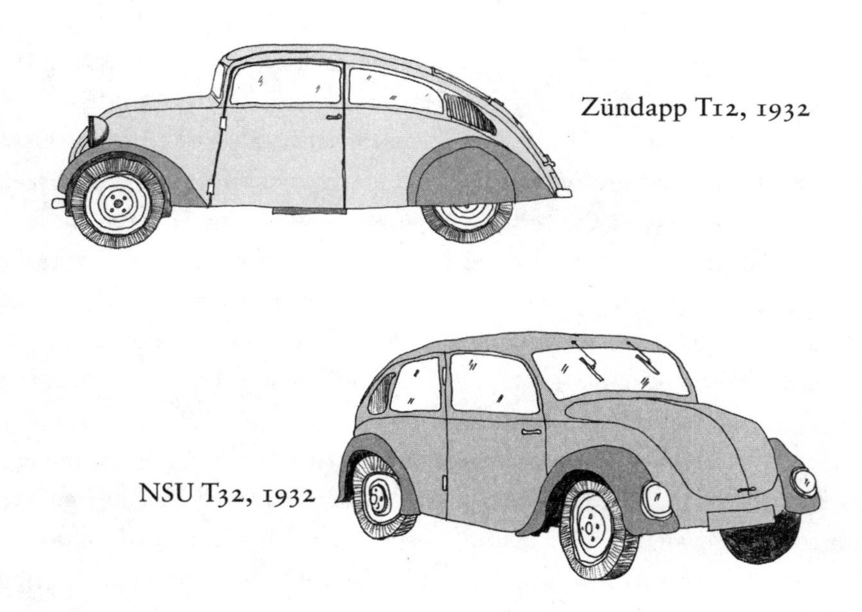

Zündapp T12, 1932

NSU T32, 1932

Later that year, NSU, a motorcycle maker that had been manufacturing cars since 1905, showed an interest in a budget car. Porsche refined the Zündapp design, giving it a curved front that complemented the sloping rear. This car, too, did not advance beyond the prototype stage. The financially strapped NSU sold its car division to Italian Fiat, which used the NSU factory to produce its own cars and shelved the Porsche design. Although these small-car projects did not come to fruition, they were an important opportunity for Porsche and his colleagues to refine the principles governing a budget car, or as a Zündapp promotional flyer had put it, *ein auto für jedermann*—a car for everyone.

ADOLF HITLER NEVER LEARNED to drive. Nevertheless, he was an avid car enthusiast and was interested in racing. At the 1933 Berlin Auto Show, when he unveiled his plan to motorize the nation with an affordable small car, he also announced a state-sponsored program to make Germany the European leader in motor racing. Porsche had not forsaken his interest in racing, and his firm had developed a design for a mid-engine racing car with a teardrop shape and a powerful supercharged sixteen-cylinder engine. Porsche had formed a partnership with the newly founded Auto Union (the future Audi), and several months after the Berlin show, he and representatives of the company met with Hitler to seek financial support from the government. With their shared interest in automobiles—and their shared Viennese background—Porsche and Hitler hit it off. Auto Union got its funding, and its racing cars, as well as those of Daimler-Benz, handily dominated the European Grand Prix circuit for the rest of the decade.* More important, Porsche established a rapport with the reich chancellor.

The leading German car manufacturers were not enthusiastic about Hitler's proposed budget car, but they felt obliged to respond to his challenge, and their trade association commissioned Porsche's firm to produce a design. Hitler had provided exacting specifications: the car should accommodate two adults and three children; it should have an air-cooled engine; it should be able to maintain a cruising speed of 100 kilometers per hour (61 miles per hour); and gas consumption should not be more than seven liters per 100 kilometers (33.6 miles per gallon). Last and not least, the selling price should not exceed a thousand Reichsmarks, about

* The German racing cars, whose aluminum bodies were unpainted to reduce weight, became known as Silver Arrows.

$400. The cheapest American car in 1935, a full-size Ford four-passenger sedan with a V8 engine, sold for less than $500, but that was in a country with a market that supported large-scale mass production. The least expensive German car—a small Opel—cost almost 1,500 Reichsmarks, so reducing that price by one-third was going to be a challenge.

Building on its previous experience with Zündapp and NSU, Porsche's team quickly produced a prototype. Ferdinand Porsche was not someone who cut corners, and thanks to his advanced engineering refinements, the final cost of the car turned out to be 1,450 Reichsmarks. The disappointed trade association called a halt to the project, but Porsche went behind its back, and in July 1936, he and his assistants drove two prototypes up to the Berghof, Hitler's Bavarian mountain retreat. Hitler loved the little cars, and on the spot he offered Porsche a contract to refine and finalize the design. Both men chose to ignore the question of the elevated selling price.

PORSCHE'S DESIGN WAS NOT the first European mass-produced budget car—the British Austin Seven preceded it by sixteen years. Herbert Austin (1866–1941), the son of a Buckinghamshire farmer, was a self-made entrepreneur with a mechanical bent. He had developed this interest while managing a factory in Australia that manufactured sheep-shearing machinery. After returning to Britain, Austin started a business making bicycles and so-called cyclecars, lightweight microcars that used motorcycle engines. In 1905, he founded the Austin Car Company, and built it into a successful business producing a variety of expensive cars. After the First World War, as the demand for luxury autos fell, and to save his failing enterprise, Austin worked at home with Stanley Edge (1903–1990), a teenage draftsman, and personally designed a budget car. There was nothing particularly novel about the Austin Seven, a two-door four-seater that was unveiled in 1922. The staid, boxy shape recalled American cars of that period, except

Austin Seven cabriolet, 1923

Tatra 11, 1923

that the Austin was much smaller—with a wheelbase of only seventy-five inches, it weighed less than eight hundred pounds. The small car had no trunk—the spare wheel was externally mounted on the rear. The water-cooled four-cylinder engine that drove the rear wheels produced seven horsepower—soon increased to ten—sufficient to achieve a top speed of fifty miles per hour, while delivering fuel economy of forty miles per gallon. The sensible "Baby Austin" proved to be a great success with the British public.* There would be many versions, including sedans, cabriolets, vans, and even two-seater roadsters, and production would continue for seventeen years.

The Austin Seven was produced under license in several countries around the world. The company producing the German version, called Dixi, was acquired by Bayerische Motoren Werke, a manufacturer of aircraft engines and motorcycles that wanted

* The Austin Seven debuted in 1922 at £155, and by 1936, the price of a sedan had dropped to £118, or about $700.

to expand into cars. In 1928, the Dixi, re-badged as the BMW 3/15, became the company's first motorcar, and remained in production for the next four years. The conservative British design did not have anything to teach Ferdinand Porsche, but this was not true of another small foreign import, the Tatra 11. The Tatra company, located in the Moravian town of Kopřivnice in Czechoslovakia, manufactured railroad cars, trucks, and automobiles. The Tatra 11 was introduced in 1923, the same year as the Austin Seven. The small four-seater had a similar shape—one box for the passenger compartment, and a smaller one for the engine—with flared fenders and a wide running board. The rear wheels were powered by a front-mounted 1,000 cc two-cylinder engine—air-cooled, so there was no radiator grille. The car was larger than the Austin Seven, weighing 1,500 pounds with a wheelbase of 103 inches, but the chief difference was the chassis, which was not a frame but a tubular backbone that enclosed the driveshaft; the differential, with independent swing half axles, was bolted to the rear of the backbone. This novel arrangement was repeated in the successor Tatra 12.

The backbone chassis was the work of Tatra's brilliant chief engineer, Hans Ledwinka (1878–1967), an Austrian who had joined Tatra in 1921, after working as chief designer at Steyr (where he was succeeded by Porsche). Following the success of the Tatra 11 and 12, Ledwinka designed the Tatra V570, an even more innovative car. For the design of the body, Ledwinka worked with Paul Jaray (1889–1974), a Hungarian engineer who had been chief of design at Luftschiffbau Zeppelin, the German airship builder. Using Zeppelin's wind tunnel, he had studied the aerodynamic performance of different automobiles and had patented numerous aerodynamic shapes. The standard measure of aerodynamic performance for a car is the drag coefficient—the lower the number, the better the performance. A boxy car such as the Austin Seven had a drag coefficient of 0.61, whereas Jaray was able to achieve drag coefficients as low as 0.23 in his wind

Tatra V570, 1933

tunnel tests. In 1923 Jaray, a Jew, left Germany and moved to Switzerland, where he established himself as a consultant, licensing his patented designs to car companies such as Tatra.

The V570 that Ledwinka designed with Jaray combined the backbone chassis with a rear-mounted air-cooled engine. The prototype body had a curved profile—front and back—integrated fenders, and an inclined windshield. These aerodynamic features enabled the car, which had a small eighteen-horsepower four-cylinder air-cooled engine, to maintain a cruising speed of fifty miles per hour. But the V570 never went into production. This was due partly to the tendency of the engine to overheat, and partly to the success of the small Tatra 57, the successor to the Tatra 11. Not wanting to cut into sales of the 57, the company decided that its next model would not be a budget car but a large sedan. Nevertheless, the little V570 prototype was influential. Hitler, who admired Tatra cars, visited Czechoslovakia and met Ledwinka. Porsche and Ledwinka were friends and frequently discussed their work, which was moving along parallel paths, as can be seen from the streamlined shapes of the Zündapp T12 and the NSU T32, which both predate the V570. Nevertheless, the similarities between the V570 and Porsche's future design for Hitler's people's car are striking.* Porsche is supposed to have remarked, "Well, sometimes Ledwinka looked over my shoulder and sometimes I looked over his."

* Tatra later sued Porsche's company for damages, claiming patent infringement, but the German invasion of Czechoslovakia made the suit moot. After the war, Tatra did win a lawsuit against Volkswagen, which settled for one million deutsche marks.

The countries of the old Austro-Hungarian Empire produced an impressive number of exceptional automotive engineers besides Porsche, Ledwinka, and Jaray. The Viennese Edmund Rumpler (1872–1940), an aeronautical and automotive engineer, built one of the first streamlined cars in 1921. Rumpler was also responsible for the swing-axle rear suspension, a key invention in the history of the automobile that enabled the drive wheels to maintain good contact on uneven roads. The Budapest-born automotive engineer Josef Ganz (1898–1967), who championed small cars in his German magazine *Motor-Kritik*, designed several microcars in the 1930s, including the Standard Superior, a two-person vehicle powered by a rear-mounted two-stroke engine. Béla Barényi (1907–1997), an ethnic Hungarian, was something of a prodigy. In 1925, when he was only eighteen and still a student, he designed a "people's car of the future," a small two-door four-seater that used a backbone chassis, a rear-mounted air-cooled engine, and a streamlined body. Barényi's prophetic student project became widely known after Ganz published the drawings in *Motor-Kritik* in 1934, and it may have influenced Porsche's design for a budget car. Barényi went on to have a long career with Daimler-Benz (see chapter 11), but Rumpler and Ganz suffered different fates. Both were Jews and were imprisoned by the Nazis in the early 1930s. Rumpler's career was ruined, and he died in 1940; Ganz fled Germany for Switzerland and eventually found his way to Australia, where he briefly worked for General Motors and died in obscurity.

Hans Nibel (1880–1934) was born in Moravia, the industrial region of the future Czechoslovakia, and after receiving an engineering degree from Munich's Technical University he joined Benz, where he rose to head the design office. Nibel was responsible for the celebrated "Blitzen Benz," a racing car that pioneered the torpedo body and held the world land speed record from 1911 to 1919. When Benz merged with Daimler, Nibel worked under Porsche. Shortly after Porsche left the firm, Daimler-Benz decided to develop a small rear-engine car, and Nibel, who was now technical director, was put in charge of the project. The result was the remarkable Mercedes-Benz 120, a small two-door four-seater that used Ledwinka's backbone chassis and independent rear suspension, combined with a small air-cooled engine. The handsome streamlined body, which predated the Tatra V570, had a rounded hood and a striking ducktail rear.

Mercedes-Benz 120, 1931

The little 120 never went beyond the working prototype stage. Like in the Tatra V570, the engine tended to overheat, a common problem with rear-mounted air-cooled engines at that time. (Three years later, Ledwinka designed an effective ducted forced-air cooling system, which Porsche would use in his budget car.) The successor to the Mercedes-Benz 120, likewise designed by Nibel, was the 130, a small rear-engine sedan, but with a water-cooled engine instead of the troublesome air-cooled model. The boxy styling was more conservative—no ducktail. Water-cooled engines were larger and heavier than the air-cooled variety, however, which unbalanced the lightweight car and adversely affected handling. The resulting modest sales caused Daimler-Benz to stick to front-mounted engines in the future. As for Nibel, in 1931 he went on to design the Mercedes-Benz 170V, a midsized car that offered the quality and features of a large Mercedes sedan (front engine and rear-wheel drive, independent suspension, hydraulic brakes) at a lower price. Sales of the popular "small Mercedes" supported the Daimler-Benz company during the difficult post-Depression years of the Weimar Republic. Nibel died in 1934 of a sudden heart attack, only fifty-four. Had he lived, he would undoubtedly have been capable of designing Hitler's people's car; after all, his

Mercedes 120 prototype incorporated all its salient features—a backbone chassis, rear air-cooled engine, and a streamlined body.

HITLER APPROVED PORSCHE'S BUDGET car design in 1936, and the next two years saw the construction of thirty prototypes (hand-built by Daimler-Benz) that were put through grueling pre-production test runs aimed at lowering maintenance and running costs. By 1938, the design was finalized. The two-door five-seater was powered by a twenty-five-horsepower four-cylinder air-cooled rear engine that could drive the car at sixty miles per hour for extended periods. The compact power plant was a so-called boxer engine, in which the cylinders were located on either side of a central crankshaft, and the pistons of each opposed pair of cylinders moved inward and outward at the same time. This type of engine had been invented more than forty years earlier by Carl Benz and had been used in the Tatra V570 and the Mercedes-Benz 120, but Porsche's car also included more recent features, such as a modified backbone chassis, which used the backbone to house the transmission, compact torsion bars instead of conventional coil or leaf springs, and four gears instead of the two or three that were common in budget cars. Heating was standard, which was also uncommon at a time when heaters were generally an optional extra. The design of the all-steel body was the responsibility of Erwin Komenda (1904–1966), an Austrian engineer whom Porsche had recruited from Steyr. The characteristic curved hood resembled the NSU T32, and the streamlined rear recalled the Zündapp T12, but both features were now smoothly integrated into an appealing bug-like form. Komenda pared away everything that was extraneous. The flowing, sinuous curves recall the earlier Art Deco style, the sculptures of Josef Lorenzl, the paintings of Tamara de Lempicka, and the glass objects of René Lalique. "Timeless classic" is a shopworn term, but in this case it applies.

Hitler had given up on the foot-dragging private car companies and determined that his car should be manufactured by

KdF-Wagen, 1938

an entirely new concern. The funding would be provided by the Deutsche Arbeitsfront (German Labor Front), which the Third Reich had created as a replacement for the banned trade unions. The Labor Front's leisure arm, called Kraft durch Freude (Strength through Joy), provided affordable vacations for working families and operated its own mountain retreats and beach resorts, even its own cruise ships. The new car was attached to this program and became known as the Kraft durch Freude Wagen, or KdF-Wagen. Linking the car to leisure was logical in a densely populated country with good public transportation—the KdF-Wagen was likely to be used chiefly for holidays and Sunday drives on the newly built autobahns. The first advertisements showed motorists picnicking and camping.

The KdF-Wagen was manufactured under Ferdinand Porsche's direction in a brand-new factory in the village of Fallersleben in Lower Saxony. The giant plant was to produce 150,000 KdF-Wagens per year, and triple that number in five years. The adjacent new company town was called Stadt des KdF-Wagens bei Fallersleben (renamed Wolfsburg in 1945). Only 630 cars were produced before war broke out and civilian car production ceased. The factory shifted to military vehicles, as well as a variety of armaments, including the notorious V-1 flying bombs. More than half of the factory workers were forced labor, chiefly from Eastern Europe.*

Although the Fallersleben factory was bombed during the war, most of the assembly line survived. There was talk of shipping the machinery abroad as war reparations, but British car manufacturers were not interested. Fallersleben was in the British sector of occupied Germany, and when the factory reopened in 1945, it was managed by the British Army and produced KdF-Wagens for the military. Eventually, production shifted to civilian models, and in 1949 the plant reverted to German management. Production of the rechristened Volkswagen increased rapidly, and by the mid-1950s the millionth VW came off the assembly line. That car incorporated such mechanical improvements as a slightly larger engine, and offered superior performance compared to the original—a top speed of almost seventy miles per hour. But, except for the quarter-glass vent windows and the oval that replaced the rear split "pretzel" window of the prewar cars, it was the same bug-like design. That was the car that took me from Hamburg to Valencia.

* In 1946, Ferdinand Porsche, who had been a member of the Nazi party, was arrested in Paris and accused of using forced labor in the KdF factory, which would make him a war criminal. He spent more than a year in French jails before being released without a trial.

CHAPTER
TWO

FOUR WHEELS UNDER AN UMBRELLA

It must be simple to maintain and inexpensive. It should be four wheels under an umbrella.

—PIERRE-JULES BOULANGER ON THE CITROËN 2CV

MY FIRST *NEW* CAR WAS A CITROËN 2CV, WHICH THE French call Deux Chevaux, or Two Horses. It wasn't actually two horsepower, although it sometimes felt that way. CV stands for *chevaux-vapeurs*, or steam horsepower, and 2CV refers to a tax category, because France, like most European countries, levies an annual road tax on car owners based on engine displacement. My car, bought in 1969 when I lived in Montreal, had a two-cylinder engine that produced twelve horsepower. That's pretty small—half as powerful as a Volkswagen and smaller than many motorcycles. It meant a top speed of about fifty-five miles per hour if you were lucky; less in a headwind. The day I bought the car, I gave a friend a ride, and driving across Mount Royal I had to keep downshifting as the car went slower and slower. Never mind, I loved my fire-engine-red Deux Chevaux. The Beatles' *Abbey Road* album was released that year, and in honor of their song "Here Comes the Sun," I painted a stylized yellow sun on the trunk lid. Well, it *was* the sixties.

What appealed to me about the Citroën? I like cars, although

I'm not particularly adept mechanically. I can change the oil and replace spark plugs or a fan belt, but that's about the limit of my competence. It was the design of the 2CV that attracted me, the inventive way that each problem was solved in a distinctive, dare I say Gallic, fashion. For example, the car rode high and had a suspension designed to produce a smooth ride over bumpy back roads. This softness produced an alarming (but quite safe) tilt when cornering, and also meant that when there were passengers in the back seat, the rear of the car sank down. If it was nighttime, that made the headlights point up instead of illuminating the roadway. *Pas de problème*: a knurled knob under the dash rotated the bar to which the headlights were attached, lowering their angle. To reduce costs and keep the doors thin, the front windows did not roll down, but had a hinged lower portion that swung *up*, enabling the driver to make hand signals (the original 1948 version had no turn indicators) and pay tolls. An openable horizontal slot beneath the windshield extended the full width of the car and provided natural ventilation—no fan required. The lightweight seats resembled lawn chairs—thinly padded cloth over rubber bands stretched between tubular-steel frames. The seats were easily detachable, and with the two rear seats removed, cargo space increased dramatically. In addition, the seats could be used as lawn chairs when picnicking or camping. The roof was fabric, because that was cheaper and lighter than metal—and the fabric could be unrolled, like a window blind. One evening, I inadvertently left the roof open and rain left an inch of water in the car; the designers had thought of that, too, and provided rubber drain plugs in the floor.

The controls were as simple as in my old Volkswagen. The speedometer and gas gauge were in a small pod behind the steering wheel, and instead of a dashboard there was a convenient shelf. The gearshift was a lever protruding over the shelf. The steering wheel was tubular steel, nothing fancy; no central horn button—instead, you beeped by pushing on the stalk that controlled the headlights. The sole luxury, installed for Canadian export mod-

Citroën 2CV, export model, 1969

els, was a large gasoline-fired heater in the engine compartment that instantaneously blasted hot air into the car.*

The Citroën car company had been founded at the end of the First World War by the industrialist André Citroën (1878–1935), an armaments manufacturer. He had early success with small, inexpensive cars, but by the 1930s, as competition increased, his company foundered. His ambitious solution was a brand-new model, a spacious four-door sedan. The Citroën 7 was popularly known as the Traction Avant, or Front-Wheel Drive. Front-wheel drive had been used by car makers ever since Lohner's Semper Vivus, although infrequently in mass-produced cars. André Lefèbvre (1894–1964), an aeronautical and automotive engineer who was in charge of the Traction Avant project, incorporated independent suspension and rack-and-pinion steering, which were also novel. The latter consisted of a circular gear (a pinion) that engaged a toothed linear gear (or rack) that converted the rotational motion of the steering column into linear motion to turn the wheels. This mechanism, introduced by BMW in the 1930s, was not only simpler and lighter than the recirculating-ball steering gearboxes it replaced, but provided the driver with a better feel for the road. The Traction Avant was intended to have a fully automatic transmission, but as development costs rose, this was changed to a three-speed manual transmission operated

* Citroën 2CVs were especially popular in francophone Quebec; the 2CV ceased to be sold in the United States in late 1960.

by a dashboard-mounted gear lever. A radical innovation was the absence of a traditional chassis. Cars typically had a structural chassis to which the body was bolted, but the welded steel body of the Traction Avant was designed as a structural shell—a monocoque—to replace the chassis, which made for a lighter, cheaper, and more crash-resistant vehicle. Lefèbvre, who raced cars as a hobby, pushed the wheels out to the four corners and located the engine well behind the front axle, which improved the balance and handling of the car.

Citroën 7,
Traction Avant, 1934

André Citroën appreciated that a car was not just a work of engineering—a car had to perform well, but it also had to look well, and that required an artistic sensibility. Lefèbvre's collaborator was a transplanted Italian sculptor, Flaminio Bertoni (1903–1964). A photograph shows Bertoni with a large plasticine model of the car, unusual at a time when European automotive design was generally done on paper. The absence of a chassis meant that the Traction Avant was about nine inches lower than a conventional sedan, and Bertoni created a low-slung body with prominent headlights, wide front fenders, and flowing lines. There is something almost menacing in the aggressive stance of the car, which would later feature in many *films noirs* and became in real life the favorite getaway vehicle of French criminals—there was even an infamous Gang des Tractions Avant.

André Citroën was a gambler, and his plan for developing and marketing an innovative car required a brand-new, retooled assembly line. The associated costs proved excessive, and by the end of 1934, despite the public accolades that accompanied the launch of the Traction Avant, his company went into receivership. Citroën's largest creditor was the tire manufacturer Michelin, which took control of the bankruptcy and eventually assumed ownership of the car maker. Pierre Michelin (1903–1937), the son of the founder, became the president of Citroën, whose new managing director and head of the engineering and design was Pierre-Jules Boulanger (1885–1950). Boulanger had an unusual background. Born in a small town in northern France, his architectural studies at the École des Beaux-Arts in Rouen were cut short by the need to work to support his parents. After completing his military service, Boulanger spent six years in the United States and Canada in a variety of jobs that included architectural drafting and running a house-building company. With the outbreak of the First World War, he returned to France and served with distinction in the Aéronautique Militaire as a navigator and aerial photographer. Joining the Michelin company after the war, he developed a close working relationship with the founder, Édouard Michelin (1859–1940), another self-made man.

With the Traction Avant in production and the Citroën company back on its feet, Boulanger saw an opportunity to expand car ownership to a new market: the half of the French population that in 1935 was still rural. The secret project was code-named TPV—Très Petite Voiture, or Very Small Car. Here is Boulanger's terse description of what he had in mind: "It must be able to carry two peasants, 50 kilograms of potatoes and a box of eggs at up to 60 kilometers per hour, and return a fuel consumption of three liters per 100 kilometers over the worst roads rural France can offer up. Furthermore, if the car were driven over a plowed field, not a single egg in the box would be broken. It must be simple to maintain and inexpensive. It should be four wheels under an umbrella." Thus, the genesis of the Deux Chevaux.

Boulanger put Lefèbvre in charge and gave him a free hand, emphasizing that it didn't matter what the car looked like, but that it should sell for a third the price of a Traction Avant. Once again, Bertoni was part of the team. Lefèbvre and Bertoni studied contemporary Czech and German budget cars—Lefèbvre met Ferdinand Porsche—but the French approach to economy would be different: they aimed to make the car as light as possible. The body was made of Duralinox, a lightweight aluminum-magnesium alloy previously used in airplane construction. The curved hood was corrugated to stiffen the thin metal, and to further reduce weight, the roof was made of a waxed cotton fabric. Bertoni's curved silhouette bore a marked resemblance to the Zündapp T12 and the Tatra V570, but he was not concerned with aerodynamics since the projected top speed was only thirty-five miles per hour. The wheelbase was the same as the KdF-Wagen, but the French car weighed six hundred pounds less, which allowed it to have a much smaller engine—375 cc versus 995 cc. The water-cooled two-cylinder boxer was linked to a three-speed gearbox. Like the Traction Avant, the car had front-wheel drive, which made for a compact engine compartment, eliminated the driveshaft, and kept the floor flat. To meet Boulanger's box of eggs challenge, Lefèbvre's team developed an innovative interlinked suspension that was connected to eight torsion bars located beneath the rear seat.

The Very Small Car was austerely functional. Boulanger encouraged Lefèbvre to be innovative, but unlike Porsche, he was strict about enforcing economies. There was only one headlight (all that French law required at the time), which was powered by a dynamo linked to the engine because there was no battery—the engine was started with a hand crank. The single windshield wiper was manually operated. The seats were canvas slings suspended from roof struts. There were no external door handles, and inside there were no instrument gauges at all. Nevertheless, Bertoni's two-dimensional curves gave the little car an unexpected sense of élan.

About a hundred preproduction prototypes were built in preparation for a launch at the 1939 Paris Motor Show, but in September of that year, France declared war on Germany, the show

Citroën
Très Petite Voiture
prototype, 1937

was canceled, and Citroën halted production of the car. Then Germany invaded, and in June of the following year, France capitulated. Boulanger, a patriot who had been awarded a Croix militaire and a Légion d'honneur in the previous war, had the prototypes destroyed lest they fall into enemy hands (a handful survived). During the occupation, he would not cooperate with the Germans and refused even to meet Ferdinand Porsche. Forced to produce trucks for the Wehrmacht, he encouraged sabotage among his workers.

Boulanger had earlier succeeded Pierre Michelin, who had died in a car accident, as head of Citroën, and after the war he restarted the TPV project. The car unveiled at the 1948 Paris Motor Show was an upgraded version of the earlier design: the boxer engine was now air-cooled with an electric starter—no more hand-cranking; the two wipers were battery-powered, as were the two headlights; the windows were glass rather than cellulose acetate plastic; the seats were mounted on the floor; and the doors had handles. The Duralinox alloy, which had proved too expensive, was replaced by steel. The Deux Chevaux was fitted with a brand-new Michelin product—radial tires—making it the first production automobile thus equipped.* The budget-minded Boulanger had relented on

* Traditional cross-ply tires were reinforced by steel or polyester cords that crisscrossed at sixty degrees to the direction of travel, while in a radial tire the plies were at right angles to the direction of travel. Radial tires provided superior road handling and better fuel economy, as well as longer life, and by the mid-1970s they had become standard on most cars.

these improvements, but he insisted that the car be available in only one color: gray.

I USED TO TAKE my Deux Chevaux in for servicing to the Citroën dealership, which was on St. Catherine Street in downtown Montreal. It was always slightly embarrassing to drive the little red car up the garage ramp and pull up beside a sleek Citroën DS—it was like parking a biplane next to a Learjet. DS didn't stand for anything—it was a pun, pronounced *déesse* in French: the goddess. *Le mot juste.*

Development of the DS had begun in the immediate postwar period, although Boulanger had been thinking of a successor to the Traction Avant even before the war. This time the code name was VGD—Voiture à Grande Diffusion, a Mass Market Car that would be larger and more technologically advanced than the Traction Avant. Once again, the team included Lefèbvre and Bertoni, as well as Paul Magès (1908–1999), a self-taught technical prodigy who had developed the unusual suspension of the 2CV. The Citroën DS was unveiled at the 1955 Paris Motor Show and was an instant hit, garnering more than 12,000 orders from prospective buyers the first day, and 80,000 deposits by the end of the ten-day show. Sadly, Boulanger was not at the launch. Five years earlier, he had been killed in an auto accident, just like his friend Pierre Michelin.

The Citroën DS had front-wheel drive, with the engine placed well back to the center of the car and the wheels pushed out to the corners. Lefèbvre designed a four-speed semiautomatic transmission (no clutch) and front-wheel disc brakes—a first for a production car. Traditional drum brakes, which were first used by Maybach in 1900, create friction by pressing a pad against the inside of a drum, whereas disc brakes squeeze a pair of pads against both sides of a disc, which is more effective both in braking and dissipating heat. Perhaps the most radical technical feature of the DS was its suspension. Postwar France had no expressways, and Boulanger was concerned that the fast-moving touring car should have a smooth ride on rough country roads. Hydraulics were commonly used in cars for braking, power steering, and the transmission, but Magès designed a hydropneumatic system that also operated the suspension, producing a smooth ride on bumpy roads. The self-leveling suspension could be controlled by the driver to increase ground clearance. I once followed a DS in a rainstorm. The road ahead was flooded, and we both stopped. The DS effortlessly raised itself up on tippy-toes and continued through the water.

Bertoni's radical teardrop-shaped body was obviously influenced by aerodynamics, and had been wind tunnel tested, but it was fundamentally a work of sculptural art. The curved windshield and rear window and integrated bumpers were features

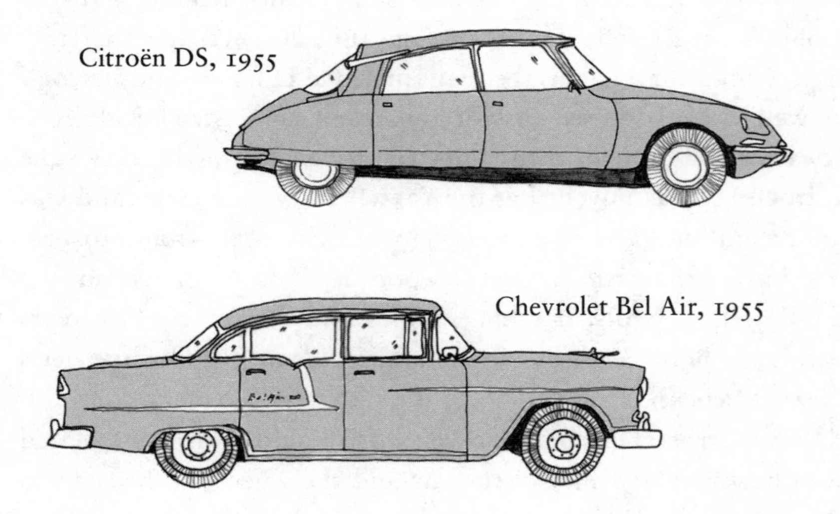

Citroën DS, 1955

Chevrolet Bel Air, 1955

that would become standard in later cars. To lower the center of gravity and improve handling, the roof was lightweight fiberglass. There were no front fenders, and the presence of rear fender skirts gave the body the sleek appearance of a speedboat. The DS was a complete departure from the Traction Avant: no aggressive stance, no masculine menace—quite the opposite. Truly a goddess.

To better appreciate the revolutionary DS, it is useful to compare it to the Chevrolet Bel Air, one of General Motors' most successful models of that same year. The two cars are roughly the same size, although the svelte DS is lighter than the boxy Bel Air (2,800 pounds versus 3,200), and consequently gets by with a considerably smaller and lighter power plant—a four-cylinder engine producing 74 horsepower, compared to a six-cylinder engine producing 123 horsepower. Top speed for both cars was 90 miles per hour. Mechanically, the comparison is between French ingenuity and American muscle, but another striking difference is the bodywork. The person in charge of styling the Bel Air was William L. Mitchell (1912–1988), who had worked as an illustrator before coming to GM and brought a graphic sensibility to the task, visible in the chrome trim and "spears" that were intended to give the impression of dynamism to the flat body panels. Chrome abounded in the Bel Air—in the hubcaps, bumpers, front grille, and hood ornament. By contrast, Bertoni used chrome sparingly, solely for window trim and for the unusual conical rear signal lights that were mounted at roof level—otherwise, the sculpted bodywork was plain. The Bel Air is often cited as an icon of the fifties, representing brash American self-confidence; the DS, on the other hand, was a symbol of *la belle France*.

My Citroën was hardly in the same league as the DS, but then it cost only about a third the price—1,500 Canadian dollars, the equivalent of about 10,000 American dollars today, making it, in 1969, the least expensive new car on the Canadian market. The cheapest new car in the United States as I write this is the Chevrolet Spark, which costs about $15,000.* The Spark is a compact hatchback, about the same size as the 2CV, but with a more powerful engine and safety features such as seat belts and air bags, as well as power steering, antilock brakes, a touch-screen infotainment system, smartphone compatibility, air-conditioning, and a rear-vision camera. In other words, the little car has standard features that were not even imagined in 1969, or were found only

* The Spark was discontinued in August 2022.

in luxury models, a reminder that automotive history is a story of rising expectations—and rising implementation—in performance, safety, and comfort.

IN 1933, EVEN BEFORE Pierre-Jules Boulanger had set the Très Petite Voiture project in motion, and the same year that Hitler proposed a "people's car," Giovanni Agnelli (1866–1945), the head of Fiat, announced that his company would develop a budget car to sell for 5,000 lire, half the price of the company's least expensive model, and roughly Hitler's projected price for the KdF-Wagen.

For some reason, the project was assigned to Fiat's aircraft engine division, where Dante Giacosa (1905–1996), a gifted young mechanical engineer, was put in charge of designing the engine and the chassis. Giacosa, a graduate of the Polytechnic University of Turin who had joined Fiat only five years earlier, began by studying the current crop of small three- and four-wheeled German minicars, such as Josef Ganz's Standard Superior. He was not impressed. "Our economy model was meant to be markedly superior to them in terms of comfort, performance, and appearance at a comparable cost of production," Giacosa wrote in his memoir. He was undoubtedly also aware of the Tatra V570 and the Mercedes-Benz 120, but knowing that Agnelli disapproved of rear-engine cars, Giacosa opted for a front-mounted engine with rear-wheel drive. He located the engine ahead of the front axle, to increase interior legroom. The extremely compact 569 cc four-cylinder engine used a thermosiphon cooling system that did not require a water pump, and a gravity-fed gas tank. The chassis consisted of two longitudinal spars, lightened with pierced holes like an airplane fuselage, and instead of simply being dead weight, the body reinforced the chassis structurally, making it a semi-monocoque. As Porsche was doing with the future KdF-Wagen, Giacosa incorporated the latest automotive technology: a four-speed synchromesh transmission, four-wheel hydraulic brakes, hydraulic shock absorbers, and independent front-wheel suspension. The unusually small fifteen-inch wheels, proportioned to the car, used tires specially made by Pirelli. The car had a top speed of fifty miles per hour.

The body of the budget car was the responsibility of Fiat's coachwork division under engineer Rodolfo Schaeffer (1893–?). Thanks to the compact engine, the curved hood sloped sharply and gave the car a distinctive profile. Schaeffer's design included cost-saving features such as sliding windows, a single windshield wiper, and rudimentary instrumentation. He dispensed with running boards. The lightweight canvas roof folded open like a Roman blind (this may have been the inspiration for the roof of the Citroën 2CV). To keep the car as light as possible, and to not compete with Fiat's next-largest small car, the ten-foot-long economy car was basically a two-seater, with a narrow space behind the seats that was large enough for two children, luggage, or, in a pinch, a single (very uncomfortable) adult. There was no trunk; the spare tire was mounted on the rear of the car.

Fiat 500 Topolino, 1936

The Fiat 500 rolled off the assembly line of the vast Fiat factory in Lingotto, outside Turin, in 1936—two years ahead of the KdF-Wagen. At ten and a half feet long, it was the smallest mass-produced car in the world. The public's nickname for the car was "Topolino," which means "little mouse" and is the Italian name for the Disney cartoon character Mickey Mouse. The round headlights did slightly resemble mouse ears, and there was definitely something cartoonish about the little car. Schaeffer's design somewhat resembled an American roadster, such as the 1933 Ford Model B, a fast car that was a favorite of John Dillinger and Clyde Barrow—the same fat fenders, prominent headlights, long hood, curved grille, and truncated rear, but endearingly miniaturized. If Mickey Mouse were robbing a bank, this is the car he would use. Of course, he would be unable to make a fast getaway, since the small engine produced only thirteen horsepower, compared to the Ford's seventy-five. The Topolino may not have been fast, but it was robust, reliable, and fuel-efficient, and it became a bestseller. The car sold for 9,750 lire in 1937, about $500—almost twice Agnelli's target, although inexpensive by European standards.

The successor to the Topolino, which appeared in 1957, was the Fiat 600, another budget car. Also designed by Giacosa, who was now the head of Fiat's Motor Vehicle Technical Office, the two-door four-seater was a step up from the Topolino. Giovanni Agnelli was no longer on the scene, and Giacosa was free to use a rear-mounted engine: four water-cooled cylinders. By then, Citroën in France and Volkswagen in Germany had had great success with their small cars, and while Giacosa was influenced by the 2CV and the rear-engined Beetle, his car was smaller than either, only ten feet, seven inches long and four feet, six inches wide. That made the Fiat 600 arguably the first so-called city car, admirably suited to the narrow streets and constricted parking of old Italian towns.

Giacosa continued to refine the design of the 600, and two years later produced what proved to be his masterpiece: a new Fiat 500 known as the Cinquecento. The wheelbase was six inches shorter than the 600's, making the car less than ten feet long. The new smaller and lighter two-cylinder engine (479 cc, versus 633) was air-cooled and produced only thirteen horsepower, yet it was enough for the light car (almost two hundred pounds lighter than its predecessor) to reach a top speed of fifty-three miles per hour. The engine was mounted transversely, parallel to the axle, which was efficient as well as space-saving. Like the early VWs, the first model didn't have a fuel gauge, only a speedometer and warning lights. The canvas roof extended all the way to the rear of the car and folded open; the lightweight front seats had tubular-steel frames. The small forward trunk compartment was barely large enough for the spare tire and a ten-gallon fuel tank, but the rear seat back folded flat to create extra cargo space. Like many early European cars, the doors were rear-hinged, which made getting in and out easier, but could be dangerous if the door whipped open while the car was moving and the passenger had to lean far out to close it; critics called them "suicide doors." Later models used front-hinged doors, which became the industry standard.

Fiat 500 Cinquecento, 1957

The bodies of the 600 and the Cinquecento, which were almost identical, were the work of the Fiat coachworks office headed by Giuseppe Alberti (1911–1978). The design is often described as "cute." The unpretentious front, with its two eyelike headlights and plump hood, gives an anthropomorphic impression to the car, which seems to be smiling. There is definitely something good-natured about this cheerful little marvel of thoughtful design. Unlike the early Tatra and Zündapp prototypes, there was no hint of streamlining; after all, the cruising speed was well under fifty miles per hour. The tiny, inexpensive Cinquecento was perfect for first-time car buyers, and over an eighteen-year production life, Fiat sold more than 3.6 million.

The Cinquecento's antecedent, the less-celebrated Fiat 600, remained in production until 1969. The slightly longer chassis and

Fiat 600 Multipla, 1956

more powerful engine proved highly adaptable, and the variety of models included small vans, pickups, upscale runabouts that resembled golf carts, and even a tuned racing version with a body from the Zagato coachworks. Perhaps the most striking adaptation was Giacosa and Alberti's 1956 transformation of the 600 into an Italian version of the VW bus, which had debuted six years earlier (see chapter 4). The body of the Multipla (Multipurpose) was a foot longer than the Fiat 600, but it used the same wheelbase, engine, and drivetrain, combined with a heavier-duty front suspension and steering assembly. The egg-shaped body had four doors and a solid roof. The sides were almost entirely glazed; wind-up windows in the front, sliders in the rear. The small spare tire (the car had twelve-inch wheels) was stored under the dashboard. The front seats were far forward, over the front axle, making the interior versatile: a removable rear bench increased cargo space, and optional jump seats in the rear increased the seating capacity to six. With the front passenger seat removed to make space for luggage, the green-and-black Multipla taxi was a common sight on Italian streets.

"HAVE YOU EVER DRIVEN a Quatr'elle?" the front-desk clerk at the car rental agency asked me. He gestured to a small light blue car standing outside. I had to admit that I hadn't—I wasn't sure what a Quatr'elle was. "Going far?" he asked. "Athens," I responded. He looked at me for a few seconds, then shook his head, gave a Gallic shrug, and went back to filling out the paperwork.

It was June 1964, and I was in Paris, embarking on an eight-week architectural road trip with a classmate. Ralph and I had met in the fourth year of our course at McGill University, worked together on class projects, and become friends. He was a Montrealer, a year older than I, and I deferred to his judgment in many matters, including the organization of the trip. Ralph had located youth hostels, looked into car rental agencies, and researched the itinerary. Our route would take us through Switzerland, across a corner of Italy, down the Dalmatian coast, inland across Yugoslavia, and south to Greece and Athens. In all, about 1,800 miles.

We loaded our bags into the back of the car, which had a large hatch. It was decided that today I would drive and Ralph would navigate. I got in the car and had a moment of confusion. I had learned to drive with a stick shift, but where was it? Not on the steering column, as in my father's Vauxhall, and not on the floor. Ah, it must be that lever with the white plastic knob that was sticking out of the dashboard. There was a large speedometer, stalk controls on each side of the steering column, one of which turned out to be the horn, and a bunch of toggle switches for lights and heating that I would figure out later. A helpful mechanic leaned in and explained the operation of the mysterious gearshift, and we were off. We first had to get out of the city, and I remember zipping up the Champs-Élysées and whizzing around the Place de l'Étoile, as it was then called. "Perhaps we should slow down a little. No need to drive like the Parisians," Ralph observed in his usual calm manner. I eased off the accelerator.

The car I was driving was a Renault 4L (pronounced *quatr'elle*). The *L* stood for "luxe," although there wasn't much luxury to be seen in the rudimentary interior—no fancy finishes, mostly painted metal, sliding instead of roll-down windows, a rubber floor mat. And who needed a radio? The hammock seats, made out of padded fabric supported by curved steel tubes, were surprisingly comfortable. So was the car's soft ride, heeling over in corners and gliding over bumps. I didn't give this much thought as we drove out of Paris, but the suspension would serve us well on the rough back roads of Yugoslavia. In fact, the Quatr'elle was a proficient off-road vehicle, and would place second in the demanding 1979 Paris–Dakar rally, right behind a Range Rover.

"Not too many frills and a workhorse" was how Ralph later described the car. The Quatr'elle was Renault's answer to the Citroën 2CV. When Pierre Dreyfus (1907–1994), the new chairman of Renault, launched the project in 1956, he described what he had in mind: "I want a versatile car, one that's urban and rural at the same time, and one that suits the needs of everyone. Call it the blue-jeans car." Like the 2CV, the Renault 4 was a four-door four-seater with a front-mounted engine driving the front wheels. The four-cylinder engine, a reliable Renault standby, produced twenty-seven horsepower and a top speed of about fifty-five miles per hour. The engine was water-cooled, but to minimize maintenance the Renault engineers had devised an ingenious sealed cooling system that never needed topping up, and unlike many cars at that time, the Renault had no lubrication points. The car had a soft and supple suspension, and rack-and-pinion steering. The roof was steel. Instead of the Citroën's rounded body, the utilitarian Renault was boxy—actually, two boxes: a big one for the people and a small one for the engine. Less character but more space. Despite being eight inches shorter than a 2CV, the boxy Renault 4 had more room, and removing the back seat created plenty of storage space. Access to the rear was by a top-hinged glazed hatch, making the Quatr'elle arguably the first hatchback. The design lacked the Deux Chevaux's quirky charm, but it had the schematic simplicity of a toy car in a children's coloring book.

The Quatr'elle had been on the market for four years when we rented it, and Renault would sell eight million of this unprepossessing little car worldwide by the end of its run in 1994, making the Renault 4 the world's number three car in terms of total production, just behind the Ford Model T and the Volkswagen Beetle.

Renault 4L, 1961

This success was due in large part to the car's adaptability and versatility. The Renault 4 wasn't a car to impress the neighbors or play boy racer. It was for transporting families, large dogs, and bags of groceries. And two wide-eyed architectural tyros.

It took Ralph and me almost three weeks to reach Athens. We were not in a hurry, and we made many stops along the way, including visits to the world's fair in Lausanne, Diocletian's Palace in Split, the medieval monasteries of Meteora, and the shrine of Delphi, the first of many ancient sites. We spent two weeks in Greece, and on the strength of reading *Zorba the Greek* I made a four-day side trip to Crete. The return journey took us across the Adriatic and up the Italian boot, with a visit to Pompeii and extended stays in Rome and Florence. The last week driving through the French Riviera was hurried, with brief stops in Cannes for a dip, and Marseille to see Le Corbusier's famous *Unité d'habitation.* We took the motorway up to Paris, arriving in the city the day before we were due to return the car. The little Renault had performed admirably, especially on mountainous backcountry roads. "Driving on dirt is a little like driving in powder snow," I noted in my journal, adding with the insouciance of youth, "I nearly drove off the road once or twice, which could be serious when there is no guard rail and a hundred-foot drop."

CHAPTER
THREE

MASS PRODUCTION

It devolves upon the United States to help motorize the world.

—WALTER P. CHRYSLER

AT THE TIME THAT ADOLF HITLER ANNOUNCED HIS GOAL of an affordable people's car, Germany lagged far behind other countries in car ownership. Whereas there was one car for every 38 Britons, and one for every 43 French citizens, there was only one car for every 134 Germans. These numbers all paled in comparison to the United States, where already in 1933 there was one car for every five Americans. The main cause of this impressive statistic was Henry Ford. Hitler, like every motor enthusiast, was aware of Ford, having avidly read the best-selling German translation of his autobiography, *My Life and Work*. In 1936, when the KdF-Wagen factory was being planned, Ferdinand Porsche was sent to America to visit Ford's vast River Rouge plant in Dearborn, Michigan, the largest factory—of any kind—in the world. A dozen years earlier, while Porsche was hand-building racing cars for DMG, and Hans Ledwinka was designing the ingenious little Tatra 11, the ten *millionth* Model T rolled off the Ford assembly line. It has been estimated that, at that time, half the cars in the world were Model Ts.

Henry Ford (1863–1947) was born and raised on a farm in Michigan, the son of an Irish immigrant father and a mother of Belgian descent. He never attended high school, and having shown mechanical aptitude, at sixteen he was apprenticed to a Detroit machinist. Exceptionally gifted, when he was twenty-four Ford constructed an internal combustion engine from scratch, and several years later built his first production car. The hand-built Quadricycle consisted of a frame with a two-cylinder engine, a chain drive, and four bicycle wheels. Ford worked for the Detroit Edison Illuminating Company at the time, and Thomas Edison, another self-taught tinkerer, encouraged the young man. Ford attracted financial backers, including the Dodge brothers, who supplied him with engines, and in 1903 he founded the Ford Motor Company. He had a flair for publicity, and the following year he personally drove a racing car he had built to set a new land speed record of ninety-one miles per hour.

When Ford started his company, there were more than 250 American car manufacturers. Although there had been attempts to build inexpensive automobiles, the most successful companies, such as Cadillac, Pierce-Arrow, Peerless, and Packard, produced hand-built luxury cars costing many thousands of dollars, at a time when the average annual industrial wage was less than $500. Ford was interested in building a mass-market car, but prudently he did not attempt to do so immediately. He named his cars alphabetically, and nineteen models preceded the Model T—although only eight of these actually went into production. They included expensive as well as moderately priced cars. The Model T would be different; in Ford's words, "it will be so low in price that no man making a good salary will be unable to own one." An ambitious goal.

The Model T debuted in 1908. The car was the work of Ford; C. Harold Wills (1878–1940), a self-taught engineer who had been with Ford from before the founding of the company; and two Hungarian engineers, Budapest-educated József Galamb (1881–1955), who had joined Ford in 1905 and became his chief designer, and Jenö Farkas (1881–1963). The front-mounted inline four-cylinder engine, which powered the rear wheels, produced twenty horsepower, sufficient to achieve a top speed of forty-two miles per hour. The two-speed transmission was controlled by a combination of clutch and hand-brake lever; a stalk on the steering column adjusted the spark timing, and a hand throttle governed the speed. The Model T included a number of mechanical innovations: the four cylinders of the engine were cast in a single block with detachable cylinder heads for easy access and repair; the crankshaft, the front axle, and the wheel spindles were made of lighter and higher-strength vanadium steel; unlike most cars, the power train was enclosed; and the thermosiphon-cooled engine did not require a water pump. The front and rear axles were mounted on semi-elliptical springs that turned the car into an effective off-road vehicle, which was much of the time, as there were few paved roads in rural America.

The *New Yorker* writer E. B. White, who owned a Model T, described the complicated process of starting the car.

> The trick was to leave the ignition switch off, proceed to the animal's head, pull the choke (which was a little wire protruding through the radiator), and give the crank two or three nonchalant upward lifts. Then, whistling as though thinking about something else, you would saunter back to the driver's cabin, turn the ignition on, return to the crank, and this time, catching it on the down stroke, give it a quick spin with plenty of That. If this procedure was followed, the engine almost always responded—first with a few scattered explosions, then with a tumultuous gunfire, which you checked by racing around to the driver's seat and retarding the throttle. Often, if the emergency brake hadn't been pulled all the way back, the car advanced on you the instant the first explosion occurred and you would hold it back by leaning your weight against it. I can still feel my old Ford nuzzling me at the curb, as though looking for an apple in my pocket.

Henry Ford, the untutored farmer's son who became a mechanic and then a business magnate, was uninterested in appearances, and the body design of the Model T is best described as serviceable. The boxy four-seater, which rode on large wooden-spoked wheels, had functional fenders, no doors in the front, a windshield but no wipers, and minimal instrumentation: an ammeter mounted on the bare wooden dashboard behind the cowl, and no speedometer. Nor was there a gas gauge; checking the fuel level required removing the front seat, unscrewing a fuel cap, and inserting a handy wooden stick into the tank. The folding canvas hood, a vestige of the horse-drawn carriage, left the sides open to the weather; in the winter, passengers simply bundled up. The first Model Ts came in four colors—gray, green, blue, and red—depending on the model; four years later, all cars were painted midnight blue with black fenders; and after 1914, when fast-drying black paint became available, Model Ts came in only that color.

Ford Model T, 1908

The Ford's functional body was less elegant than the 1901 Mercedes 35HP, but people did not buy a "Tin Lizzie," as the car came to be called, for its looks. The original $825 Model T was not the cheapest car available, but it was the most dependable, easiest to operate, and easiest to repair. Henry Ford conceived of the Model T as essentially a farmer's car—city dwellers could travel in streetcars and omnibuses. America was still predominantly rural, and

Ford understood that in a country as large as the United States, where distances were great, the automobile would have particular appeal to the residents of isolated farms and small towns. The robust construction and simple mechanics were geared to rural users, and the car's success was a function of its versatility. Its ground clearance and suspension allowed it to navigate rough gravel roads with ease. The heavy (1,200 pounds) vehicle had a powerful engine, and fitted with aftermarket studded rear wheels, it could serve as a tractor. With one wheel removed and the addition of a belt drive, the car turned into a stationary power plant for threshers, water pumps, wood saws, and electric generators.

At a time when many wealthy car owners traveled with their own mechanics, the rugged Model T was easy to maintain and modify by the do-it-yourselfer. The car was also designed to be easy to manufacture. Ford did not invent mass production—there were already mass-produced American products such as Mason jars, Ames shovels, Eli Terry clocks, and Overman bicycles. Nor was he the first to use an assembly line—that honor belongs to Oliver Evans, who built a waterwheel-powered flour mill in Philadelphia, and to the Swiss inventor Johann Bodmer, who built a number of factories in England in which he combined stationary workers with conveyance systems. But Ford was the first to apply the idea to automobile manufacturing when, five years after he built his factory, he had the novel idea of moving partially completed car bodies from workstation to workstation. This single change eventually reduced the assembly time of a car from 12.5 man-hours to a remarkable 1.5 man-hours. By 1921, Ford had lowered the selling price of the Model T to $325, the equivalent of four months of average earnings, about five thousand present-day dollars. That made the car forty percent cheaper than its closest competitor. The Ford Motor Company built factories in Canada and England, and assembly plants in half a dozen European countries as well as Mexico, Brazil, Argentina, and Japan. Over the nineteen-year life span of the Model T, more than fifteen million were built.

Henry Ford understood that mass production depended on mass consumption. A benevolent dictator (he was opposed to unions and held many reactionary views), over the years, he increased wages and reduced work hours, and in 1926 announced that henceforth his factories would be closed all day on Saturday, creating a full two-day weekend. More time to go for a drive! What Ford did not anticipate was that mass demand would be inconstant. Although over the years he made improvements to the Model T, such as adding a six-volt battery and an electric starter, Ford resisted making cosmetic alterations. He stuck to his utilitarian philosophy of extreme standardization on the assumption that there was no need to change a well-functioning and well-priced product. He was wrong. The same buyers who were excited by the Model T in 1908 had become blasé a decade later. They grew bored with Ford's insistence on standardization, and demanded variety, which Ford's competitors, especially Chevrolet, were happy to provide. After more than a decade of lively growth, sales of the Model T stagnated. Ford had come face to face with a fundamental paradox of mass production: standardized production stimulates individual consumer demand, which can only be met by non-standardized products.

WALTER P. CHRYSLER (1875–1940), a dozen years younger than Henry Ford, was likewise a self-taught machinist from the Midwest. Like Ford, he began with innate technical skills and subsequently developed exceptional abilities as a manager and entrepreneur. Chrysler started in the railroad industry, and ended up successfully managing a factory in Pennsylvania for the American Locomotive Company, the second-largest steam locomotive builder in the country. He was fascinated by automobiles; as a teenager he built a car with a two-cylinder engine of his own design, and when he was thirty-three he bought his first car, an expensive Locomobile. Before driving it, he took the engine entirely apart—several times—to understand how it worked. Three years later, he was recruited by General Motors as the works manager of its faltering Buick division. Thanks to his meticulous reorganization, Buick became General Motors' most profitable division, and Chrysler ended up as president of Buick and a vice president of the largest automobile manufacturer in the world.

Chrysler was now independently wealthy, and in 1919, finding himself at odds with the leadership of General Motors, he decided to retire. He was only forty-four and still interested in cars, so when he was approached by a consortium of banks to run Willys-Overland, an established car company that was on the verge of bankruptcy, he accepted, demanding—and receiving—the astonishing annual salary of $1 million. He later took a similar position with the ailing Maxwell Motor Corporation, which provided Chrysler with the opportunity to start his own company. His idea was to build a compact, maneuverable automobile with the quality and performance of a large luxury car, but at a midlevel price. He enlisted a trio of singular engineers. Fred Zeder (1886–1951), Owen R. Skelton (1886–1969), and Carl Breer (1883–1970) had met while working for Studebaker and had formed their own automotive engineering consultancy. All were university graduates in mechanical engineering. Zeder was the manager, Skelton the designer, and Breer the analytical whiz; Chrysler referred to them as the Three Musketeers.

The result of Zeder, Skelton, and Breer's work—with input from Chrysler—was the Chrysler Six, unveiled at the New York Automobile Show in 1924. The car came in a variety of bodies: coupe, sedan, convertible, town car, and an eye-catching roadster. For the sedan, they adopted the popular two-box design that would serve car designers for years: one box for the passengers, and another smaller one for the engine. Unlike early automobiles such as the Model T, which tended to have "open bodies" with retractable canvas roofs, the Chrysler Six adopted the new "closed body" concept in which the passengers sat in an enclosed cabin with a roof and large windows.

The spare tire was mounted on the rear; although there was no place for luggage, in later models an actual steamer trunk was mounted on the rear, giving rise to the name still used for the luggage compartment.

In 1924, American car companies did not have styling departments. They either depended on independent coachbuilders who provided design services and supplied the bodies, or on an in-house "body engineer." This person was not an artist like Flaminio Bertoni at Citroën; he was usually a mechanical engineer. At Chrysler, Oliver H. Clark, a junior member of the Zeder-Skelton-Breer office, was responsible for coordinating the bodywork, much of which used existing Maxwell parts. A rare styling feature of the Chrysler Six was a large chromed radiator grille with a cap topped by a winged Viking helmet hood ornament. The curved front and rear fenders were integrated with a wide running board. While not breaking any new ground, with its squarish proportions and prominent headlights, the boxy car had character.

Chrysler 6
Series 70 sedan,
1924

The Chrysler Six was one of the first production cars to have a dashboard and a full set of instruments: a speedometer, an odometer and trip meter, ammeter, oil pressure gauge, water temperature, and fuel gauge. Unlike the Model T, the main controls—brake, clutch, and accelerator/throttle pedals—were on the floor, as was the stick gearshift. This arrangement would become an industry standard. Mark Howell, an early automobile historian, called the Chrysler Six the "first modern car" and recounted his teenage experience of being behind the wheel.

> On starting the engine, I was struck by the uncanny absence of those sounds so common to others. No clicks from the valve gear; no whine from the camshaft drive. Just a comforting tautness, as though each part was perfectly shaped to fulfill its function. The engine seemed to run with a freedom that suggested total absence of friction. The controls were especially light and precise in action. Touch the brake pedal, and the perfectly equalized hydraulic system responded immediately. Touch the throttle, and response was so instant as to suggest eagerness. Even gear shifting had been transformed from heavy drudgery to an act of swiftness and ease. Delightful! But these sensations rapidly faded once under way, for here was a quality of performance startling in its contrast with the conventional. It was as though one had driven for the first time free of dragging brakes and retarded spark.

Zeder, Skelton, and Breer had worked hard to create this smooth experience. They designed a compact six-cylinder high-compression engine that put out an impressive sixty-nine horsepower, more than all but one of its competitors. The top speed of the car was seventy-five miles per hour, at a time when sixty was considered fast. Mechanical refinements included a seven-bearing crankshaft, which increased stability and reduced vibrations, and four-wheel hydraulic brakes, at a time when many cars had brakes only on the rear wheels. Novel rubber engine mountings dampened vibrations. The three-thousand-pound car handled well and drove smoothly thanks to its use of "balloon" tires, which were thicker and wider than conventional tires and were previously available only on such luxury cars as the Pierce-Arrow sedan, which cost $6,000, compared to $1,195 for a Chrysler Six. Thanks to all these features, Chrysler sold an impressive (for a brand-new marque) 32,000 cars in the first year, 100,000 the second year, and 182,000 the third. By 1927, thanks largely to the Six, the new Chrysler Corporation—only two years old—had become the fourth-largest automobile company in the United States, after General Motors, Ford, and Hudson. By 1928, having acquired Dodge Brothers, Chrysler was producing 360,000 cars a year and, having overtaken Hudson, had become part of what people were starting to call the Big Three.

IN 1927, HENRY FORD, now in his sixties, reluctantly agreed with his son Edsel (1893–1943), who was the nominal president of the company, that they should finally replace the venerable Model T. The result was the Model A.* The Model A was bigger than the Model T—longer, wider, almost twice as heavy, with a larger four-cylinder engine (3.3 liters versus 2.9), producing twice the horsepower and a greater top speed (sixty-five miles per hour, versus forty-two). The mechanical improvements included an unsynchronized three-speed transmission, hydraulic shock absorbers, and four-wheel drum brakes. The Model A continued the mechanical simplicity of the Model T; the gas tank was mounted behind and above the engine, eliminating the need for a pump. The redesigned controls consisted of a floor stick shift, and brake, clutch, and accelerator pedals. The instrument pod included a speedometer, odometer, ammeter, and gas gauge.

Styling of the Model A was the responsibility of Edsel Ford and József Galamb. For the entry-level Tudor sedan, a two-door four-seater, they adopted the practical two-box design that was becoming an industry standard. Like the Chrysler Six, the Model A had a rear-mounted spare tire and no trunk. The front windshield, which was made of laminated safety glass (a first for the American car industry), was top-hinged to provide ventilation.

* Models W, X, and Y were manufactured in England. The new model simply restarted the alphabet.

The interior was more basic than the Chrysler Six, as befitted an entry-level car.

The Model A Tudor was just one of seven body styles that included a four- and two-door sedan, a coupe, a sports coupe, a roadster, a phaeton, and a small truck. The cars came in four colors: Niagara Blue, Arabian Sand, Dawn Gray, and Gun Metal Blue. Combining the reliability and simplicity of the Model T with features that were normally found on more expensive cars such as the Chrysler Six, the Model A was a commercial success. During its relatively short production run—1928 to 1932—almost five million were sold.

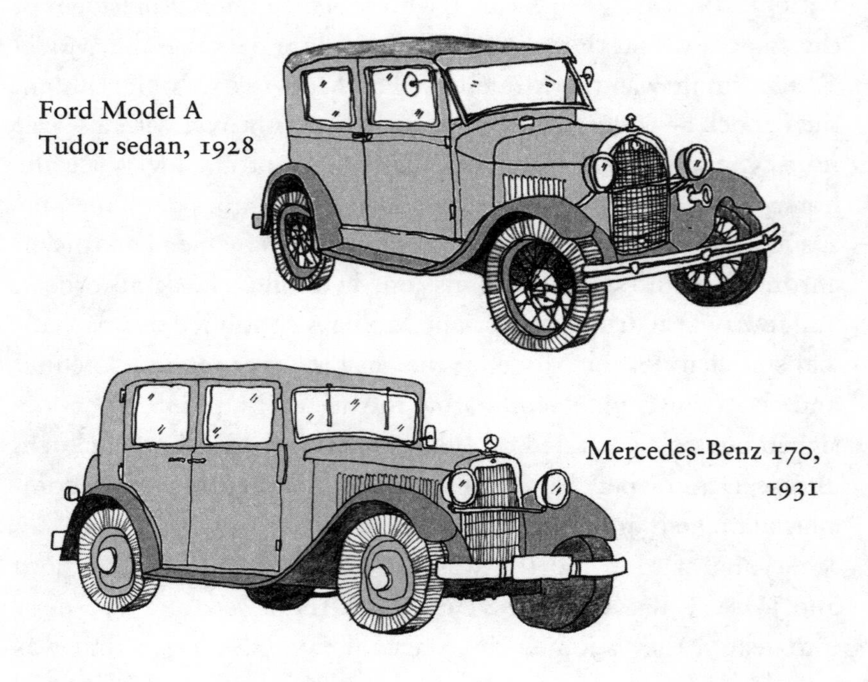

Ford Model A Tudor sedan, 1928

Mercedes-Benz 170, 1931

Five million in five years! It is the scale of production that set American cars apart from their European counterparts. For example, the five-year production run of the successful "small Mercedes" that Hans Nibel designed in 1931 amounted to fewer than 14,000. The Mercedes-Benz 170 sedan and the Ford Model A Tudor were comparable; both were front-engine, rear-drive, two-box designs with an identical one-hundred-inch wheelbase. The four-cylinder Ford engine was more powerful than the six-cylinder Mercedes (forty horsepower versus thirty-two), and the Ford's top speed was sixty-five miles per hour, versus fifty-six for the German car. Both had four-wheel drum brakes, and both came in two- and four-door versions, although the largely hand-built Mercedes-Benz had a higher-quality finish. But it was mass production and the assembly line that made the real difference: the Ford cost $500, compared to the equivalent of $950 for the Mercedes.

IN THE EARLY DAYS of the American automobile industry, most car manufacturers relied on coachbuilders to design and build car bodies. Ever since the Model T, the Briggs Manufacturing Company had been a major supplier of bodies to the Ford Motor Company. The Briggs design department was headed by John Tjaarda (1897–1962), a Dutch-born aeronautical engineer who had worked for General Motors. Tjaarda, who had served in the Dutch Air Force, brought his European experience to the work, and, inspired by the example of Paul Jaray, he tested car models in a wind tunnel to improve aerodynamic performance. In 1933, with Edsel Ford's encouragement, Tjaarda designed a streamlined show car—the Briggs Dream Car—a mock-up of which was prominently displayed in the rotunda of the Ford Motor Company's pavilion at the Century of Progress Exhibition in Chicago. The Dream Car was a low-slung teardrop design with a unibody—which is what Americans called a monocoque—prominent fenders, and the aggressive stance of a Traction Avant. The unusual headlights emerged from the fenders, and the two rear portholes were so small that the car needed a periscope-style rearview mirror. The roomy interior had three-abreast bench seats whose tubular-steel supports made them resemble modernist sofas. Although the six-seater Dream Car was much larger than Porsche's KdF-Wagen, its rear engine and sloped front and rear anticipated that design.

Briggs Dream Car, 1933

John Tjaarda was not the only Detroit automotive engineer interested in aerodynamics. Carl Breer, too, had started to experiment with a small wind tunnel of his own construction. He was encouraged by Walter Chrysler, who was interested in design, having recently commissioned the striking Art Deco Chrysler Building in Manhattan. The Chrysler Corporation built a large wind tunnel where Breer could test carved wooden models of various contemporary cars, such as the Chrysler Six and the Model A. Breer determined that these boxy designs performed extremely poorly aerodynamically, and his solution, as Paul Jaray had also found, was to make the front curved and the rear sloped to reduce turbulence.

Breer, Skelton, and Zeder set to work to develop an aerodynamic car. They used a conventional front-engine and rear-wheel-drive configuration, but included several engineering innovations. The body was a partial-unibody design, all steel except for the roof.* The rear seat was located ahead of the rear axle, which improved comfort for the passengers and distributed the weight more evenly, resulting in a smoother ride. The eight-cylinder engine produced an impressive top speed of ninety miles per hour. The ten-foot wheelbase and the width of the car resulted in a cav-

* Like all cars at that time, the Airflow roof was a wood frame covered in fabric. It was not until 1934 that U.S. Steel produced sheet metal in eighty-inch widths, wide enough to cover an entire car roof.

ernous interior with three-abreast seating, front and rear. This was before cars were air-conditioned, and to improve ventilation, the panes of the split windshield could be tilted open, the front window and quarter-glass vent could be opened separately or rolled down as a unit, and the rear seats had quarter-glass vents as well as roll-down windows. There was space for luggage behind the rear seat, although oddly, no openable trunk lid; the spare tire was mounted externally on the rear. Oliver Clark, the body engineer, translated the results of the wind tunnel tests into sheet metal, resulting in a teardrop shape with a rounded front and a sloping back. To increase aerodynamic performance, the headlights were integrated into the body and the split windshield was given a slight V-shape. The drag coefficient of the car was 0.5, which is comparable to that of the future Volkswagen Beetle (0.48).

Chrysler Airflow CU, 1934

The Chrysler Airflow was launched in 1934. Despite its mid-range price of $1,345, superior performance, ingenious features, and generous amenities, the car was not a commercial success—in the first year, it sold fewer than 10,000. There were several reasons for the poor sales. Chrysler, afraid that its competitors might launch a similar streamlined car, had rushed production, resulting in numerous defects as well as delays and canceled orders, which had hurt the Airflow's reputation. But the main reason was that buyers did not warm to the car. "Born in a Wind Tunnel" was the theme of the marketing campaign, but aerodynamics and lower wind resistance were rather vague concepts, especially at a time of powerful engines and inexpensive gas. The usual American luxury car of that period had a long hood, an impressive vertical grille, exaggerated fenders, and massive headlights; the snub-nosed Chrysler Airflow had none of these. The car made sense from an engineering point of view, but Zeder, Skelton, and Breer needed an artist like Bertoni to pull things together aesthetically, as he did with the Traction Avant, which debuted the same year. The Airflow, with its blunt front, odd grille, and watered-down Art Deco touches, not only lacked glamour, but to many it seemed downright homely. The following year, Chrysler hired the industrial designer Norman Bel Geddes (1893–1958), who was a vocal proponent of streamlining, to redo the front, but it was too late. Sales continued to decline, and after four years Walter Chrysler withdrew the car from production. It was his first—and only—business misfire. To add insult to injury, Paul Jaray, who had filed American patents for his aerodynamic improvements, sued Chrysler for infringement, and the car maker was obliged to pay a licensing fee.

Perhaps Chrysler should have bought Jaray on board the Airflow team, as Ledwinka did with the contemporaneous Tatra 77, a remarkable car that was unveiled to great acclaim at the 1934 Paris Motor Show. Ledwinka and Jaray scaled up the principles that they had developed in the small Tatra V570 prototype—a rounded front and a sloping rear. The Tatra 77 was a large luxury sedan, a spacious four-door six-seater with four-wheel independent suspension and a rear-mounted air-cooled V8 engine. An unusual feature of early models was that, to improve visibility, the driver sat in the middle of the bench seat, flanked by passengers.

The body design of the Tatra was the work of Erich Übelacker (1899–1977), a talented German automotive engineer and a graduate of Prague Technical University. He had joined the company in 1927, and worked on the Tatra 57, as well as all the later streamlined models. Like the Airflow, the Tatra 77 had a curved front and a sloping rear, but an important difference was that

Tatra 77, 1934

the Czech car was twelve inches lower, which made for a sleek profile. The rear engine was located beneath the steeply sloping back, which was bisected by an unusual dorsal fin that divided the airflow (there was no rear window). Luggage was stored in a large space behind the rear seats; the spare tire and the gas tank were in the front. The aerodynamic design included an underside that was entirely shrouded in metal to reduce wind resistance. The Tatra used a lightweight magnesium alloy for the engine, transmission, suspension, and body, which made it about seven hundred pounds lighter than the Airflow. Thanks to its light weight and aerodynamic features, the three-liter engine, which produced only 56 horsepower, achieved a top speed of 87 miles per hour (the Airflow needed a five-liter, 117-horsepower engine to produce a similar speed). The Czech car had an extremely low drag coefficient of 0.38, which would only be equaled two decades later by the sleek Citroën DS. The Tatra 77 and its successor, the Tatra 87, were widely perceived as groundbreaking vehicles, valued not only for their performance but also for their striking appearance, which reinforced the impression of effortless speed. These cars were hand-built in relatively small numbers and were high-priced—the Tatra 87 cost a third more than the most expensive Airflow.

The Chrysler Airflow debacle put a damper on innovation at Chrysler, but it did not discourage Edsel Ford, who had remained interested in aerodynamics ever since the Briggs Dream Car. He convinced his father that Lincoln, Ford's luxury division, which Edsel had been instrumental in acquiring, needed a smaller and sleeker model than the Lincoln Model K, a massive car that often served as a limousine. Edsel Ford engaged John Tjaarda to collaborate with Eugene T. "Bob" Gregorie (1908–2002), a young designer who had worked for yacht builders as well as General Motors and whom Edsel had hired and recently promoted to head Ford's design department. The result was the 1936 Lincoln-Zephyr, which incorporated several features from the Briggs Dream Car: a modified unibody construction, skirts over the rear wheels, a raked windshield, integrated fenders with built-in headlights, and a fastback—that is, a smoothly sloped rear. Like all American cars, the Lincoln had a front-mounted engine driving the rear wheels.* The engine was a flathead V12, putting out 110 horsepower and a top speed of ninety miles per hour. The basic four-door sedan sold for $1,320.

Lincoln-Zephyr, 1936

The dimensions of the Lincoln-Zephyr were similar to the Airflow, although the Lincoln weighed less than the somewhat overengineered Chrysler. Some of the aerodynamic features were similar, too—the integrated headlights, raked windshield, and fastback—yet the overall impression was different. Partly, it was a matter of more harmonious proportions and a simpler design—no fussy quarter-glass windows. Briefly put, it was a handsome car. Instead of the Airflow's curved hood with its odd "waterfall"

* American car makers favored front engines. The sole exceptions were the short-lived Tucker 48 and the 1960 Chevrolet Corvair (see chapter 6).

grille, the Lincoln-Zephyr had an elegant split grille and a prow like the bow of a speedboat.

A 1937 promotional film made by Ford shows a pair of businessmen on board a TWA flight. One of them has been working out of the country, and the other is telling him about various technological advances in transportation. When the plane lands, a lady friend of the returning expatriate pulls up on the tarmac (those were the days!) in a Lincoln-Zephyr. "Now, *there's* an unusual-looking car," the expat says. "What kind of car is this?" "That's the latest example of automotive engineering," the other explains, "and a very good example of that streamlining I was telling you about." A porter opens the trunk, flips down the spare tire, and loads several suitcases. As the car pulls away, the camera lingers on the parked DC-3. The message is clear: a streamlined plane and a streamlined car. While the Lincoln-Zephyr was not as aerodynamic as the Tatra 77, it had a decent drag coefficient of 0.45. The Zephyr sold well: 15,000 cars the first year (compared to 10,000 Airflows), and the model became the mainstay of the Lincoln division.

Ford Model 68 coupe, 1936

The basic Ford workhorses in the mid-1930s were the Model 48 and its successor, the Model 68. This platform was used for roadsters, sedans, coupes, and convertibles, as well as a station wagon and a pickup truck; the four-door sedan sold for $575. Ford had introduced an affordable flathead V8 engine three years earlier, and thanks to that powerful motor, Ford outsold its rival

Chevrolet that year, with more than 800,000 units. While Tjaarda did not incorporate radical aerodynamic concepts into the cars, he rounded off the boxy edges and made the body smoother and sleeker, a sculptural approach that would dominate American car design until the 1950s.

Only twenty-seven years had passed since the introduction of the Ford Model T, and the mass-produced American automobile had experienced a radical evolution, from utilitarian all-purpose machine to consumer product. Each year brought improvements in performance, ease of operation, and comfort, improvements so regular that they came to be taken for granted by car buyers. These tangible upgrades were accompanied by less tangible changes in appearance. Cars were becoming stylish artifacts to be appreciated visually, coveted, admired, even cherished.

CHAPTER
FOUR

GOOD TOOLS

It does everything. It goes everywhere. It's as faithful as a dog, as strong as a mule, and as agile as a goat. It constantly carries twice what it was designed for, and still keeps on going.

—ERNIE PYLE ON THE ARMY JEEP

I'VE ONLY OWNED ONE AMERICAN CAR. IN 1985, MY WIFE and I were living on a farm in southern Quebec, and we needed a car that could pull a trailer and transport bushel baskets of apples from our orchard to the wholesaler's warehouse. Our house was at the end of a thousand-foot driveway that accumulated drifts in the winter, so we also needed something that could push through drifting snow. There was a Chevrolet-Buick-GMC dealership in our village, and I dropped in to take a look. There was a vehicle on the lot that seemed perfect: a functional-looking, no-frills two-door truck that resembled what would later be called a sport-utility vehicle. I liked the practical black vinyl interior and the red exterior—which I later discovered was actually called Apple Red. Someone had ordered the truck and changed their mind, and the dealer wanted to sell. I wanted to buy, and we made a deal.

Our vehicle was a GMC S-15 Jimmy. GMC is the truck division of General Motors, and S-15 referred to the popular compact quarter-ton pickup truck that was the platform of the Jimmy, which

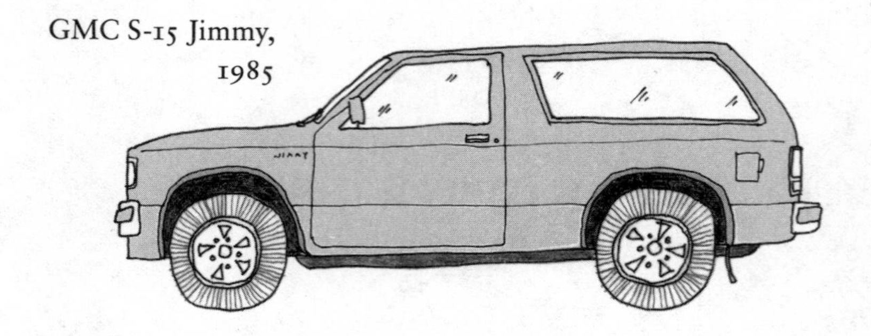
GMC S-15 Jimmy, 1985

had the same two-door cab, combined with an enclosed body that replaced the open cargo bed. The configuration reminded me of the Renault that Ralph and I had rented in Paris decades earlier. The Jimmy was three times larger (4,056 pounds, compared with 1,323), but it had the same functional simplicity: one box for the people, another for the engine. The rear hatch was a split gate consisting of a frameless glass section that swung up and a tailgate that swung down. The plain design had the straightforward functionality of a hand tool: it was meant to do one job—in this case, moving people and stuff—and with the back seat folded, that meant a lot of stuff.

GMC called the Jimmy "a truck you can live with." The compact S-15 was not an off-road vehicle; ours had rear-wheel drive, powered by a two-liter four-cylinder engine—not fast, which was just as well, since it handled like a truck. The experience of driving was different from our previous cars, though not for that reason. First of all, you sat up high. The suspension made for a soft ride, which gave the impression of floating. The interior was roomy—much larger than a Renault 4 or a Deux Chevaux. And driving the Jimmy, I felt anonymous, just one of the crowd—keep on truckin'.

Our Jimmy was a distant relative of the station wagon. Station wagons had emerged in the early 1900s, almost at the same time as the car itself. Custom coachbuilders, using a Model T chassis, had added a wooden body equipped with benches, a canvas roof, and open sides. These so-called depot hacks were used to ferry travelers—and their luggage—from suburban railroad stations. Seeing a business opportunity, in 1923 Durant Motors manufactured a wood-bodied "station wagon," and in 1929 Ford added a station wagon to its Model A lineup. The boxy vehicle carried eight passengers in three rows: 3-2-3. The rear cushioned bench seats—"blue-gray artificial Spanish leather"—were removable, and the lowered tailgate was sturdy enough to support a steamer trunk. Matching side curtains could be put up in rainy weather. A Ford advertisement read, "The new Ford Station Wagon has been designed to meet the needs of large estates, country clubs and families having summer homes in the country or by the seashore. It is particularly well suited to such

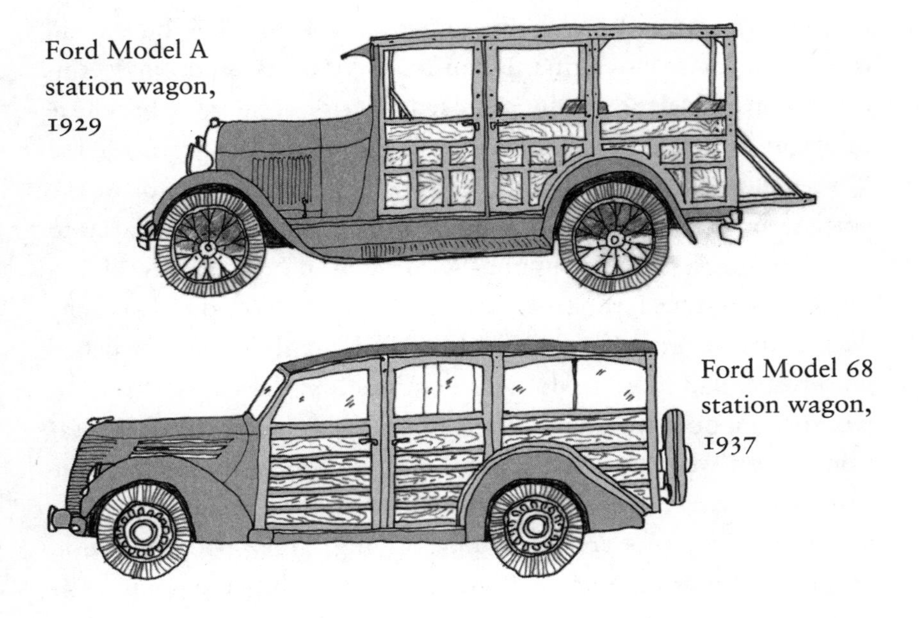

Ford Model A station wagon, 1929

Ford Model 68 station wagon, 1937

use because it combines the sturdiness of a light truck with the flexibility and comfort of a passenger car." The accompanying illustration showed the car parked not at a railroad station but on a ferry dock in the stylish summer getaway of Mount Desert Island, Maine, surrounded by well-dressed summer visitors in cloche hats and straw boaters.

As the ad suggested, the station wagon was intended for the well-to-do. The Ford wagon cost $795, which was sixty percent more than a Model A sedan, a hefty premium for what was essentially an open-air vehicle. The maple-and-plywood body of the Model A wagon was manufactured for Ford by the Briggs Company. The successor Model B wagon had a similar boxy design, but with a larger engine and a wooden body that was manufactured by Ford itself, using timber from its own forestland in Michigan's Upper Peninsula. This station wagon was the most expensive car in the 1933 lineup, and of the more than 300,000 Fords sold that year, fewer than 2,000 were station wagons.

Ford was the leading manufacturer of station wagons in the 1930s. By 1937, the Model 68 wagon had a more car-like profile, with integrated headlights and a sleek front. The buyer had the option of a V8 engine, a third row of seats, and glass windows that made for a fully enclosed interior. The sales brochure underlined the station wagon's social cachet: "For transporting a weekend party to the yacht, to the stables, the lodge, or to the cottage by the seashore, it is ideal since it enables the party to travel en masse, taking supplies with them in comfort." Whereas most cars emphasized streamlining, the boxy "woodies," as they were called, were different. The beautifully crafted and varnished woodwork had more to do with boatbuilding than cars. Not that wood was a practical material for a car. It was heavy (the Ford station wagon weighed almost two hundred pounds more than the standard sedan), the wood expanded and contracted in the sun and rain, the joints creaked while driving, and varnished wood required regular repainting. But in a way, that was the point. Like tennis whites and yachting apparel, the woodie did not belong to the everyday world.

Another ancestor of our Jimmy appeared in 1935. Chevrolet had outstripped Ford in total sales, but it did not have a station wagon, and when the company finally decided to add one to its lineup, Chevrolet engineers took a different approach. Instead of putting a wooden body on a car chassis, they modified an existing half-ton panel delivery van, adding two windows on each side and installing seats that accommodated eight passengers in three rows (3-2-3). There were several advantages to this approach: the load capacity of a van was greater than that of a car; the body was all steel, which was more practical than wood; and using existing body parts from the van reduced the selling price—$580—which was dramatically lower than that of a Ford woodie.

Chevrolet Suburban Carryall, 1935

The low price was important because the utilitarian Suburban Carryall was aimed at a different clientele. Chevrolet called the car an "all-purpose vehicle" in its upbeat advertisements. "First—it's a rugged, fast delivery unit of the de luxe Station Wagon type, with modern load, storage and handling conveniences. Second—it's a trim sedan with spacious, comfortable interior seating eight persons." In other words, the Carryall was conceived as a practical tool for the self-employed carpenter, plumber, or electrician; a work car that could double as a family car. The "de luxe" interior was actually rather plain: bare metal, a dashboard that included only basic instrumentation, and rear windows that did not open. A heater, radio, second windshield wiper, even a rear bumper were extras. Like its panel truck progenitor, the car had only two doors, and a tailgate and tilt-up window in the back. The two rear bench seats were removable, leaving a cavernous fifty-two-by-seventy-five-inch open floor space.

Over the years, the Suburban (the Carryall name was soon

Chevrolet Suburban, 2009

dropped) increased in size—and cost—and gained creature comforts, a finished interior, rear doors, a larger engine, and four-wheel drive. But it never lost its all-purpose function. Despite being affected by changing car styles—by 2009 it had a raked-back windshield—the Suburban preserved its truck-like appearance, becoming the longest-lived nameplate in automobile history. *Car and Driver* called the Suburban a "quintessentially American vehicle . . . big, brash, potent, and pragmatic." A favorite of the Secret Service, the car has appeared in so many blockbuster action movies that it has its own star on the Hollywood Walk of Fame. Suburbans transported politicians and celebrities, ranchers and emergency workers, drug lords and suburban moms. And its small relative, the S-15 Jimmy, carried a pair of tired apple pickers and bushel baskets of McIntosh apples.

CARS LIKE THE SUBURBAN took automobile design in a different direction, and so did another working vehicle: the jeep. The jeep originated in the mid-1930s, when the US Army decided that, as part of its drive to motorize, it needed a small, quarter-ton scout car. The British military had already experimented with such a car based on an Austin Seven, and Austin's American subsidiary had provided the US Army with trial models. American Austin had since been transformed into the locally owned American Bantam Car Company, which manufactured what today would be called subcompact cars. In consultation with American Bantam, the US Army developed performance specifications for a small multipurpose scout car that could double as a field ambulance and a machine gun carrier. The army's requirements included: three bucket seats, on-demand four-wheel drive, a payload of six hundred pounds, a top speed of fifty miles per hour, a maximum wheelbase of seventy-five inches, a maximum track width of forty-seven inches, and a maximum height of forty inches (with the windshield folded). In addition, for purposes of airborne operations, the weight could not exceed 1,300 pounds.* In 1940, with the European war already underway, the army sent a request for proposals to 135 car companies. The request included a demanding delivery schedule: eleven days for the initial proposal; forty-nine days to produce a working prototype; and seventy-five days to deliver seventy test vehicles. This was perhaps *too* demanding, for only two companies responded: American Bantam and Willys-Overland. Willys had experienced ups and downs since Walter Chrysler had run it, and was currently producing the Willys 77, a short-wheelbase sedan that sold for less than $500 and which the company hoped could form the basis for the scout car.

American Bantam submitted detailed blueprints and cost estimates and committed to producing a working prototype in seven weeks, while Willys-Overland asked for an extension, and the army awarded the contract to American Bantam. The team was led by Harold Crist, an experienced automotive engineer who had worked for Duesenberg and Stutz. American Bantam's two-seater roadster, which had the specified seventy-five-inch wheelbase, served as the starting point for the new design. The American Bantam engineers used existing stampings and parts and substituted a forty-horsepower engine for the standard American Bantam twenty-horsepower model. Karl Probst (1883–1963), a freelance engineer, prepared a set of drawings. The open scout car had no doors, fenders only on the front, a fold-down wind-

* The weight limit was later raised to 2,600 pounds, and alternative lightweight models were developed for airborne operations.

shield, and utilitarian squarish proportions with a blunt front and recessed headlights. Oddly, while not styled, the result was stylish in a bare-bones sort of way.

In September 1940, Crist drove a completed prototype vehicle from the company factory in Pennsylvania to an army vehicle test center in Maryland, and delivered it as promised. At this point, things got complicated. The military brass was satisfied with the American Bantam Reconnaissance Car (BRC), but was concerned that the small company, which currently produced fewer than a thousand cars a year, would not be capable of manufacturing hundreds of thousands of vehicles. As a result, the competition was reopened. Willys-Overland and the Ford Motor Company, which had somehow become involved in the project, were invited to submit designs, in addition to American Bantam. The little company, which was strapped financially and needed the work, was hardly in a position to object.

The army provided Ford and Willys with the American Bantam blueprints, so not surprisingly the three prototypes were remarkably similar. The army, needing to move forward, accepted all three proposals and ordered 1,500 preproduction vehicles from each company for use in additional trials, as well as to be exported to Britain and the Soviet Union under the Lend-Lease program. Following the trials, in July 1941 an order for 16,000 vehicles was awarded to Willys-Overland. The Willys MB was not only cheaper than American Bantam's BRC ($739 versus $1,166), but its durable fifty-five-horsepower engine was more powerful. When it became clear that even Willys would have trouble meeting production quotas, the second order of 30,000 vehicles was given to Ford, although the army insisted that Ford adhere to the Willys design. American Bantam, which had been responsible for the original concept, was out of the picture. Over the course of the war, Willys produced about 360,000 vehicles and Ford about 280,000.

"I don't think we could continue the war without the jeep," the war correspondent Ernie Pyle wrote in a 1943 newspaper column.

American Bantam
BRC, 1940

Willys-Overland
MB, 1941

"It does everything. It goes everywhere. It's as faithful as a dog, as strong as a mule, and as agile as a goat. It constantly carries twice what it was designed for, and still keeps on going." The source of the term *jeep* is unclear. It was likely a blend of First World War military slang, a character in a 1935 *Popeye* comic strip called Eugene the Jeep, and a derivation of GP, or General Purpose, which was Ford's name for the car. What is certain is that, as early as 1941, *jeep* shows up in newspaper reports, army publications, and Willys advertisements.

With the war drawing to a close, Willys-Overland developed a civilian version of the jeep aimed at farmers, ranchers, and industrial users. The Jeep CJ (Civilian Jeep), which went into production in 1945, had the same engine, on-demand four-wheel drive, a slightly heavier transmission, lower gearing, and an identical body, with the addition of a tailgate and a drawbar for towing. The basic model came with only a driver's seat, and among the many extras were additional seats, a canvas top, a heater, front or rear power takeoff, and heavy-duty springs. Willys-Overland called it a four-function vehicle: a runabout capable of speeds up to sixty miles per hour; a tractor with a 1,200-pound drawbar; a truck with an 800-pound payload, and a mobile power unit. In other words, a modern Model T. A *cheerful* Model T—the first CJs were green with yellow wheels or tan with red wheels, and other color combinations were added later. In the first four years of production, Willys sold more than 200,000 CJs. This success inspired several competitors, including the International Harvester Scout and the Ford Bronco, which had enclosed bodies, and the Chevrolet K5 Blazer, which was built on a full-size pickup truck chassis.

A British version of the jeep emerged in the immediate postwar period. The Rover Company was an established firm founded in 1877 by John Kemp Starley (1855–1901), the inventor of the Rover Safety Bicycle, the first modern bicycle. The company moved on to motorcycles and in 1904 expanded to automobiles, and by the 1920s was known for its high-end sedans. During the war, Rover manufactured aircraft and tank engines, but after the war it found itself in a quandary. There was little demand for expensive cars, which were highly taxed, and in any case, rationing limited new car production. The managing director of Rover was Spencer Wilks (1891–1971), whose brother Maurice (1904–1963) was the chief engineer. Maurice Wilks owned a farm in North Wales, where he kept a US Army surplus jeep, which is what gave him the idea. The Willys-Overland Jeep CJ was already in production in the United States, but it was not exported to Britain. What if Rover manufactured a comparable car? As an agricultural and commercial vehicle, a British jeep would be exempt from luxury taxes, and if Rover exported it to Commonwealth countries, the company would be exempt from material rationing. At this point, the Wilks brothers imagined the project as purely a stopgap measure until the British economy recovered.

The vehicle that Maurice Wilks designed resembled an American jeep—which was not surprising, since the first prototype used a Willys-Overland MB chassis. The Rover had a similar eighty-inch wheelbase, a comparable four-cylinder engine, on-demand

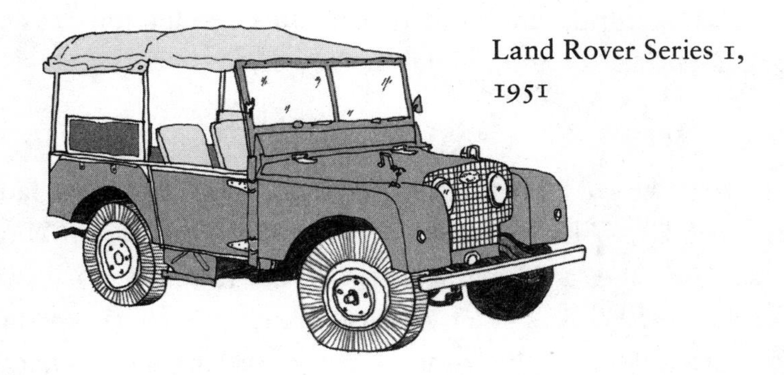

Land Rover Series 1, 1951

four-wheel drive, an open body with a tailgate, an optional canvas roof, and optional front and rear power takeoffs. There were important differences, however. Although using the same forty-eight-inch track dimension as Willys, the body was wider and could seat three abreast, which meant that the car could carry seven passengers, or as many as ten in the long-wheelbase version. The chassis was a sturdy box-welded frame. The body panels were either flat or made with two-dimensional curves to simplify production, and the car had doors. Because steel was in short supply, an aluminum alloy was substituted for the body panels. Lighter and noncorrosive, this proved so effective that even when steel became available, Land Rover bodies continued to be made of aluminum. The vehicle was available only in sage green, the color of a wartime surplus paint originally intended for airplane cockpits.

The Land Rover, as the car was called, debuted in 1951, a basic vehicle that sold for the equivalent of $665 in 1948 (compared to $1,090 for a Jeep CJ). The Land Rover excelled in many different settings: as a farm truck, as a military vehicle, and as overland transportation, equally at home in the Australian outback or on a country estate. The simple, almost improvised design, with its distinctive, no-nonsense character, was oddly appealing. The vehicle that was intended as a stopgap measure turned out to be a runaway success, and annual production was soon 25,000, doubling in the following decade, and by the time production ceased in 1985, more than 1.3 million had been sold.

MEANWHILE, IN THE UNITED STATES, a year after the launch of the Jeep CJ, Willys-Overland unveiled the Jeep Station Wagon. The car used the same engine as the Jeep CJ, but in most ways it was different. Willys's chief engineer, Barney Roos (1888–1960), who had worked on the original army jeep engine, designed an independent front suspension using a transverse seven-leaf spring, and extended the wheelbase from eighty inches to 103. The body was the work of Brooks Stevens (1911–1986). Stevens belonged to a new generation of freelance car stylists who called themselves industrial designers and included Raymond Loewy, Norman Bel Geddes, and Henry Dreyfuss. They designed home appliances, office furniture, and automobiles, as well as railroad trains, airplane interiors, and even consumer packaging—Stevens was responsible for the Miller Brewing logo. He gave the Willys station wagon a Jeep-like front, with a steel grille, round headlights, and large fenders. The company bought its car bodies from third-party coachbuilders, and Stevens designed panels that could be easily fabricated by any competent metal shop. The result was functional, if not exactly beautiful. The spacious interior had more headroom than most station wagons and could sit six—or nine, when fitted with a rear-facing third bench. All the passenger seats were removable to increase cargo space. The two-door wagon came with rear doors or a tailgate. The all-steel body and the general configuration were a smaller version of the Chevrolet Suburban Carryall, except that the Willys rode and handled more comfortably than the heavier Suburban. It was a truck-like car rather than a car-like truck.

The Jeep Station Wagon had rear-wheel drive, but in 1949 an on-demand four-wheel drive option became available. In addition to the standard floor-mounted gearshift, a second lever sent power to the front wheels, and a third lever engaged a low gear for

Willys-Overland Jeep Station Wagon, 1946

extra pulling power. This was a historic moment: while the open Jeep CJ had been the first mass-produced civilian car with four-wheel drive, the Jeep Station Wagon was a fully enclosed family car that was also a capable off-road vehicle. Sometimes, the car was advertised as a "Utility Wagon," aimed at industry and business; sometimes, it was called a "Sedan Wagon," perfect for holiday outings. One senses hesitancy in the marketing, as if Willys knew that it had produced something important, but was not quite sure what to call it.

Willys-Overland referred to its jeep-like station wagon as a "go-anywhere car" and a "common sense car that leads a double life," and priced it at a modest $1,241, which was $40 less than Chevrolet's Carryall and $300 less than a full-size Ford station wagon. But selling a double life proved challenging, especially when the station wagons manufactured by the Big Three were flashy sedans stretched into wagon shapes, adorned with chrome and decorated with varnished wood. The Jeep Station Wagon in a four-wheel-drive version had all the characteristics of a later sport-utility vehicle, but it would be another thirty-five years before car buyers fully embraced that concept.

The successor to the Jeep Station Wagon appeared in 1963. Kaiser Motors, which had acquired Willys-Overland, unveiled the Jeep Wagoneer, a vehicle that sat six and had a large cargo area accessed by a rear liftgate. The Wagoneer stuck to the idea of a go-anywhere car, but combined it with something unexpected:

luxury. The car sold for over $3,000, the price of a high-end sedan, and included similar features: independent front suspension, power steering, automatic transmission, and optional air-conditioning, as well as on-demand four-wheel drive. The engine was a flathead six-cylinder putting out 140 horsepower, more than twice as powerful as the Jeep Station Wagon. Likewise designed by Brooks Stevens, the Wagoneer was ten inches longer than its predecessor, but because it was eight inches lower, it was more car-like, both in appearance and handling. Stevens's approach, which *Popular Science* magazine called "country-club styling," was distinctly conservative, low-key rather than flashy, to appeal to upper-middle-class buyers. He used the same two-box configuration as the earlier car, but there was nothing utilitarian about the interior, which was finished like an expensive sedan.

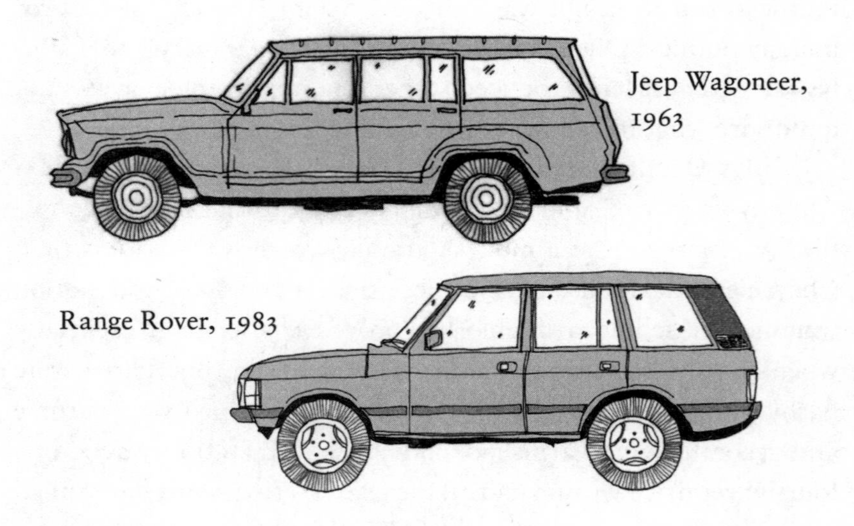

Jeep Wagoneer, 1963

Range Rover, 1983

A comparable full-size four-wheel-drive vehicle appeared about seven years later in Britain. Spurred by the success of the Ford Bronco and the International Harvester Scout, in 1970 Rover, which had earlier made several attempts to develop a larger version of the Land Rover, launched the Range Rover. Like the Bronco, which the Rover engineers had studied closely, the new car was a boxy vehicle with a rear liftgate, and a 100-inch wheelbase that made for a roomy interior. The 127-horsepower V8 engine produced a top speed of 96 miles per hour, making for an exceptionally fast road car as well as a capable four-wheel-drive off-roader. The design of the aluminum body was straightforward—an engineer's aesthetic—although the carefully crafted tall car with large windows had that indefinable quality: chic. Unlike the Wagoneer, the Range Rover was not intended to be a luxury car; in its original form, it had only two doors, a utilitarian plastic dashboard, and a vinyl interior designed to be hosed down. Then something unexpected happened. The car was taken up by the urban beau monde—a London wit dubbed it the "Chelsea tractor." During the 1980s, as London boomed, the chic Range Rover moved upmarket and became available with air-conditioning, power steering, an automatic transmission, alloy wheels, and a carpeted interior with leather seats and walnut inlays. With the accoutrements of a luxury vehicle, a well-equipped four-door version cost as much as a Porsche 911.

The Wagoneer and the Range Rover were out of reach of most car buyers, but they established the four-wheel-drive utility wagon as a prestigious vehicle, even a status symbol. Both cars had serious off-road capabilities, which was part of their attraction, even if most drivers did not actually venture into the rough. They were like another status symbol of the 1980s, the Rolex watch; it was nice to know you could take it 660 feet below the ocean's surface, even if all it really had to endure was an occasional dip in the pool.

An affordable version of the four-wheel-drive utility wagon appeared in 1984. The American Motors Corporation (AMC), which had acquired Jeep from Kaiser, launched the Jeep Cherokee XJ, a midsize car that was smaller than the Wagoneer. The Cherokee, which met new federal fuel economy standards, was 1,200 pounds lighter than the bulky Wagoneer, and had a small four-cylinder engine. The body design was the responsibility of Richard A. Teague (1923–1991), who was in charge of styling at AMC.

The innovative suspension was the work of Royston Lunn (1925–2017), a British-born engineer who was head of engineering for Jeep and had previously worked for Ford, where he was responsible for the GT40 racing car. The Jeep Cherokee had a rigid body welded to a frame—the first four-wheel-drive vehicle built this way. The car sat five; the large cargo area was accessed by a one-piece liftgate that was made of fiberglass to make it lighter and easier to operate. Like the Range Rover, the crisp design conveyed stylish, no-nonsense utility.

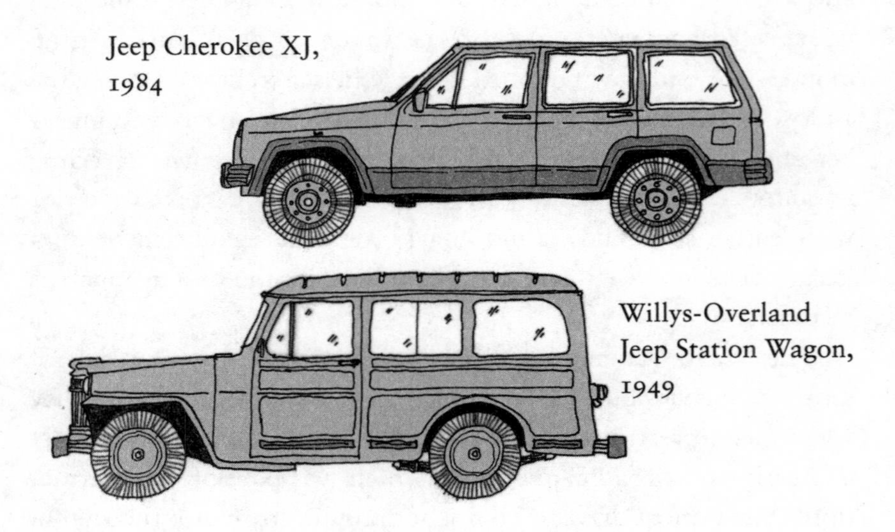

Jeep Cherokee XJ, 1984

Willys-Overland Jeep Station Wagon, 1949

It is worth comparing the Cherokee XJ to the original 1949 Willys-Overland four-wheel-drive station wagon. Both were two-box designs. Both had four-cylinder engines, although the high-compression XJ engine put out 110 horsepower, compared to sixty-three for the older car. The Cherokee had none of the features that the 1949 model inherited from the woodie and the army jeep; no paneling, no squarish fenders. The Cherokee's sloping windshield and slightly sloping rear suggested a degree of streamlining (although the drag coefficient was a relatively high 0.52). The XJ was almost a foot lower than the Willys-Overland, which improved handling and also heightened the impression that it was a car rather than a truck. The AMC team included Renault engineers, the two companies having made a cooperative agreement to develop and market cars. I like to imagine that some of the genetic material from the little Renault 4 seeped into the design process.

In 1984, the Cherokee XJ cost about $10,000 (compared to $18,000 for a Wagoneer) and proved extremely popular; during its eighteen-year production run, almost three million were sold. AMC called the car a SportWagon. The schizophrenic name did not stick—or, at least, only in part. Increasingly, people were referring to tallish four-wheel-drive wagons like the Cherokee and the Wagoneer as "sport-utility vehicles," or SUVs. They came in a variety of sizes, from compact to full-size, and in a wide price range, from economy to luxury.

Car-like utility vehicles had been around since the 1935 Chevrolet Suburban Carryall, and the changes that propelled the SUV were societal rather than automotive. First among these was the growing popularity of rural vacation homes, and associated outdoor recreation activities such as hiking, skiing, camping, kayaking, and biking, for which the go-anywhere SUV seemed admirably suited: the car had plenty of cargo space, and plenty of ability to handle the Great Outdoors. Not just to handle, but to be *seen* to handle: "I may be commuting to my desk job, but I'd rather be spelunking" was the message. Second was affluence. More American families could now afford two cars, and rather than having to choose between a four-by-four utility vehicle and a traditional sedan, they could have both. Third was the change in where Americans lived. Despite the name of the Chevy Suburban Carryall, in the 1930s most Americans still lived either in large cities or small towns. The postwar suburban housing boom of the 1950s changed that. Suburbs depended on personal rather than mass transportation; that is, on cars. The comfortable and easy-driving SUV, with good visibility and plenty of room for passengers and cargo, suited the suburban lifestyle.

The Jeep Cherokee was not the only automobile that appeared

in 1984. That same year, Chrysler introduced a radically different suburban vehicle: the Dodge Caravan and its Plymouth sibling, the Voyager. The concept of this car was simple: a roomy box—what automotive designers called a "monospace"—with a large sliding side door on the nearside for convenient access. The Caravan was three feet shorter than a full-size station wagon, yet with an optional third row of seats, it could seat up to seven passengers, and all were facing forward, unlike station wagons, in which the third row usually faced the rear. The Chrysler van was fifteen inches lower than a full-size commercial van, so it could fit into a suburban garage, hence the name: minivan.

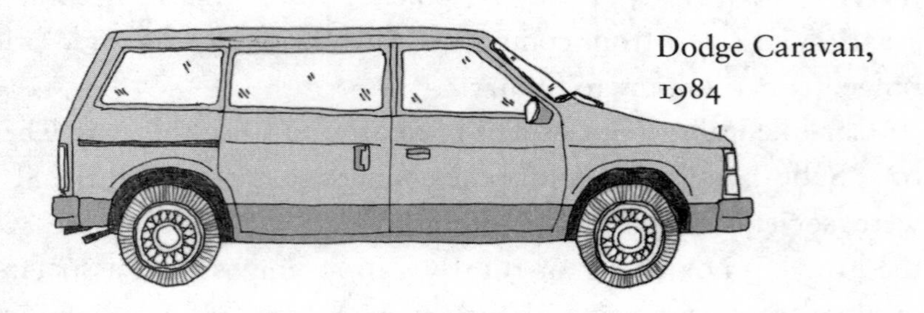

Dodge Caravan, 1984

Although the Chrysler minivan was a van, it used a car platform, and because it was front-wheel drive, the floor in the passenger compartment was perfectly flat, yielding a taller interior than a station wagon, as well as excellent visibility for the driver, who was seated in a tall captain's chair. The compact 2.2-liter engine was not powerful, but it was dependable and delivered good gas mileage. This was neither a truck nor an off-road SUV—it drove like a car. There was nothing sporty about the minivan; convenience rather than performance was the byword—a car for ferrying the kids to school or soccer practice, picking up groceries from the supermarket, and shopping at the mall. *Car and Driver* called it a "mom-mobile."

The docile minivan was the brainchild of the new head of Chrysler, Lee Iacocca (1924–2019), and his chief product planner, Harold K. Sperlich (b. 1929). Iacocca and Sperlich had worked together at Ford on the Mustang (see chapter 8), and both had been fired by Henry Ford II. One of their unrealized Ford projects was the Mini-Max, a front-wheel-drive people mover that was the genesis of the minivan. The minivan was a gamble for Chrysler, which was on the verge of bankruptcy, recently bailed out by a federal government loan. The gamble paid off, and the Caravan/Voyager was a success, spawning imitations by Ford and General Motors as well as by Japanese manufacturers.*

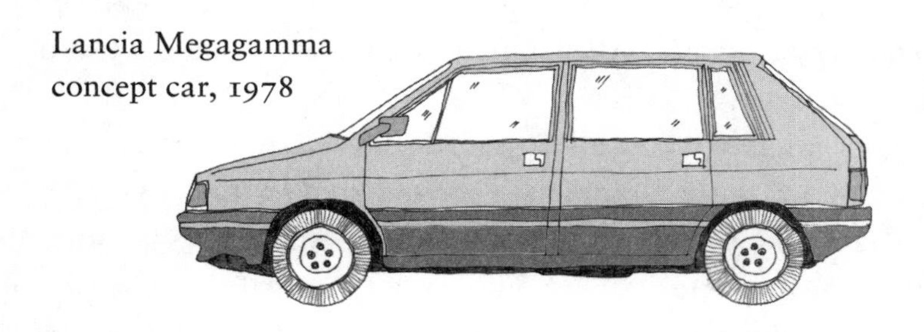
Lancia Megagamma concept car, 1978

The minivan idea was hardly original; after all, the Fiat 600 Multipla dated from 1956. A more recent predecessor was a concept car that Lancia had developed in 1978. The Megagamma was designed by Giorgetto Giugiaro (b. 1938), a celebrated car designer whose studio in Turin, Italdesign Giugiaro, was responsible for exotic Italian roadsters, upmarket German sedans, and Korean family cars, as well as the best-selling successor to the Beetle, the VW Golf. The Megagamma was a four-door, two-box car with a front engine, front-wheel drive, and a tall, roomy cabin. The design was based on a car the Museum of Modern Art had commissioned from Giugiaro and Alfa Romeo for a 1976 exhibit

* The minivan displaced the station wagon; production of the two remaining full-size American station wagons ceased in 1996. (Station wagons continue to be popular in Europe, especially Germany.) By 2000, minivan sales had hit their peak, and in 2020 the Dodge Caravan was discontinued, having been in turn displaced by the SUV, which had come to dominate the market. By 2020, more than half of all new car sales in the United States were SUVs.

on a future New York taxicab. That car was even more minivan-like, since it had a sliding passenger door.

But the actual progenitor of the minivan dated back to the 1940s. In 1947, Bernardus Pon (1904–1968), a Dutch car dealer who was the first Volkswagen dealer outside Germany, was visiting the Wolfsburg factory where he saw a flatbed transporter built on a converted VW chassis. This gave him the idea of a delivery van, which he proposed to VW management. Pon's concept was a boxy body mounted on a standard VW rear-engine chassis, with the driver sitting up at the front, above the axle, and the rest of the space available for cargo.

Volkswagen was still under British Army control, and it was not until the company reverted to German management in 1949 that the minivan took shape. Although the so-called Type 2 Transporter turned out to require a sturdier chassis than the passenger car, the wheelbase remained the same, as did the

Volkswagen Type 2 Transporter, 1950

small air-cooled four-cylinder rear engine. The rear portal axle, a two-gear reduction that increased torque, was adapted from the Kübelwagen, a Porsche-designed scout car that was made in the KdF plant during the war. After wind tunnel tests, the boxy Transporter was given a split windscreen and curved corners, which resulted in a surprisingly low drag coefficient of 0.44. The almost square flat front had none of the usual characteristics of

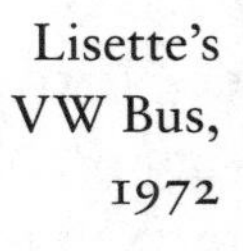

Lisette's VW Bus, 1972

an automobile—no radiator grille, no hood, no fenders, just two large headlights, like two lemur eyes.

Production began in 1950. The Transporter was available in two models: a panel truck and a passenger vehicle with side windows and a removable bench seat. A version of the latter, seating nine and called the Microbus, appeared the following year.* Because the engine blocked the rear, access was provided by a pair of hinged doors on the nearside. In 1964, a sliding door was introduced as an option, and three years later became a standard feature. Thus the VW bus was born.

Working with third-party coachbuilders, Volkswagen marketed a camper version that turned the bus into a miniature motor home with a collapsible bed, folding table, cupboard, icebox, and sink—not as palatial as an American Winnebago, but a step up from a tent. Many people built their own. This was especially common in North America in the 1960s and '70s, when the boxy Volkswagen bus was adopted by the so-called counterculture, partly because of its capacity and practicality, and partly as a thumb in the eye of convention. Not your mother's motor home.

I never owned a Volkswagen bus, but I once converted one into a camper. My friend Lisette was planning a long road trip from Montreal to points south, and she had bought a bare-bones VW

* The VW Microbus was called a *kombinationskraftwagen* (multiuse vehicle), or *kombi* for short, and that term is still used to mean "station wagon" in Germany, as well as in Poland, Hungary, Romania, and Sweden.

bus—no rear benches, just two seats up front. I offered to complete the interior. I had always been interested in the design of sailboat cabins, and this seemed like a similar sort of design problem. The interior was fifty-six inches high; you had to stoop to move about, but there was plenty of headroom once you sat down. The open space behind the driver's seat was nine feet long and five feet wide, the main constraint being the engine compartment, which was a sort of low box at the end. The commercial VW camper had an elaborate unfolding sofa bed that extended over the engine compartment, but that was too complicated for my limited carpentry skills. I designed a bench that served as a settee and that hinged open to become a twin bed. The bed was supported by two storage bins, one for clothes, and the other—insulated—for food. The space over the engine compartment served as a counter, with storage drawers beneath. It was all pretty basic, just varnished birch plywood. An inside-out woodie.

CHAPTER
FIVE

CAR CULTURE

My primary purpose for twenty-eight years has been to lengthen and lower the American automobile, at times in reality and always at least in appearance.

—HARLEY J. EARL

MORE THAN ONCE, IN READING ABOUT THE HISTORY OF the American automobile, I came across mention of a Golden Age, usually referring to the 1950s. It strikes me that the 1920s and 1930s were more innovative in terms of car design, but the fifties were golden in at least one sense: ownership. Half of all American families owned at least one car, and by the end of the decade that fraction had risen to three-quarters.

Although most of the pioneering inventions of the motor age originated in Europe, mass production and mass ownership of cars arrived first in the United States, and it was here that the car took hold. In a large country of great distances and spread-out towns and cities, individual mobility had great appeal, just as Henry Ford foresaw. But it was more than that. The automobile suited Americans. Mobility suited their sense of individual freedom and choice—no need to rely on the vicissitudes of public transportation. The car suited the American fascination with mechanical devices, not to mention gadgets, and it also appealed to the national inclination for tinkering—a flourishing aftermarket in do-it-yourself accessories and car parts emerged together with the Model T. In addition, automobiles such as the van and the pickup truck were a boon to the self-employed. The landscape historian J. B. Jackson has written of the importance of commercial vehicles. "They are hauling small loads, sometimes hauling passengers, but they are also making repairs, installing and removing and replacing and servicing small businesses." For all these reasons, automobiles assumed a central position in American life.

"It devolves upon the United States to motorize the world," Walter Chrysler had announced as early as 1928. Car ownership would increase in Europe and eventually everywhere—he was right about that—but nowhere would it have the widespread effect on everyday life that it did in the United States. Americans' relationship to the automobile has often been described as a love affair, but as the historian John Lukacs wisely observed, it was really more like a marriage. And, as in a marriage, the union changed both parties. The car, which began as a utilitarian device, turned into a recreation, a hobby, a status symbol, and more. The car culture produced motor courts, motels, and drive-in restaurants, drive-through windows and fast food, drive-through banks, and even drive-in churches. In 1932, a New Jersey entrepreneur received a patent for a drive-in movie theater—the invention described a series of ramped parking places that provided unobstructed lines of sight to the large screen; by the 1950s, there were more than four thousand drive-ins across the country. Cars enabled people to travel to summer cottages, far-off ski slopes, and national parks. Families took Sunday drives, teenagers worked on jalopies and hot rods, crowds flocked to speedways and drag races. Thumbing or hitchhiking became popular—Jack Kerouac's transcontinental epic *On the Road* appeared in 1957. Football stadium parking lots spawned tailgate parties. The automobile changed American life in profound ways. Car commuting killed off the evening newspaper—you couldn't read while driving—but by the end of the 1950s more than half of all cars had radios and talk radio soon emerged. Cars also radically transformed the urban landscape. New interstate highways facilitated travel to the outer fringes of the metropolis and hastened the demise of interurban streetcars, railroad travel, and traditional downtowns, which were replaced by suburban strips. Shopping centers, service stations, car washes, and used car lots proliferated in this decentralized landscape.

Americans liked to keep their cars close, and in the newly built suburban communities, attached carports and garages were now an integral part of the home. The car in the garage was usually large and heavy, with a powerful engine. Unlike in Europe, there was no annual tax on engine displacement; instead, road taxes were included in gasoline prices, and gasoline was cheap. Americans used their cars for long trips, so there was an emphasis on comfort, and features such as a soft suspension, power-assisted steering and brakes, and powered windows and seats were standard on many models, even on an entry-level car such as the Chevrolet Bel Air. By the 1950s, high-end cars such as Cadillacs, Lincoln Continentals, and Imperials had air-conditioning.

Part of the car culture was brand loyalty; there were Ford families and Chevy families. Many car owners regularly traded in their cars for new models, or, if they were moving up in the world, they might trade up from a Chevy to a Pontiac, or from a Ford to a Mercury. The man largely responsible for the idea of trading up was Alfred P. Sloan Jr., the head of General Motors. Sloan (1875–1966) was born in New Haven, Connecticut, and graduated from the Massachusetts Institute of Technology with a degree in electrical engineering. After the roller bearing company he owned and operated was acquired by General Motors, he rose in the corporation and in 1923 became president of what was then an ailing enterprise. Thanks to his leadership, GM recovered, and by the 1930s it had overtaken Ford as the largest car company in the world. One of Sloan's pioneering concepts was the creation of brands that represented increasingly expensive cars, from the entry-level Chevrolet to the top-of-the-line Cadillac. The brands did not compete with each other and appealed to different income categories of car buyers. Equally important, the price range—from $500 to $3,500—offered plenty of opportunities for trading up.

In 1924, the five GM brands were Chevrolet, Oldsmobile, Oakland (predecessor of Pontiac), Buick, and Cadillac. Sloan determined that GM needed an additional luxury brand, to fill a gap between Buick and Cadillac. The result was the flamboyant LaSalle, which attracted more than 25,000 buyers during its first two years.* The car was available as a sedan, coupe, convertible, roadster, and phaeton. Smaller than a Cadillac, but sharing the same build quality, the LaSalle 303 also shared the same powerful V8 engine, making it sporty and fast. Above all, with a low torpedo body, the 303 *looked* fast. Its long hood, sleek fenders and running boards, wire wheels, and massive headlights gave it a sense of continental flair that suggested a widely admired European car such as the French-built Hispano-Suiza. The Hispano-Suiza cost $15,000, while a four-door LaSalle phaeton cost $2,500, a thousand less than the staid Cadillac. This price was possible because of the way the car was built. The LaSalle was manufactured and marketed by Cadillac, but unlike the Cadillac and the Hispano-Suiza, which were largely hand-built, the LaSalle rolled off an assembly line, just like a Chevrolet.

As American automotive technology improved, it became difficult to differentiate models purely on the basis of performance, but as the LaSalle demonstrated, buyers responded to innovations in prestige and beauty—that is, to *design*. The man responsible for the bodywork of the LaSalle 303 was Harley J. Earl (1893–1969). Born in Hollywood, Earl learned automotive body design while working for his father, a Los Angeles coachbuilder who specialized in custom car bodies, many built for movie stars. The Earl Automobile Works was bought by the West Coast Cadillac distributor, who put the young Harley in charge of custom bodywork. That is how Earl came to the attention of the head of the Cadillac division, who hired him to over-

* The LaSalle, caught short by the Depression, did not have time to establish itself as a luxury brand, and, outsold by its competitor Packard, it was discontinued in 1940.

LaSalle Model 303 Phaeton, 1927

see the styling of the new LaSalle. Sloan was impressed by the success of the LaSalle, and in 1927 he established a new design department at GM and put Earl, who would become his protégé and friend, in charge. The fifty-person Art and Color Section was a new idea in an industry where bodywork was traditionally considered the prerogative of mechanical engineers, or, as we have seen, of outside body shops. Earl recruited a staff with diverse backgrounds: art, graphics, advertising, coachbuilding, and industrial design. Despite initial internal resistance to what detractors called the "Beauty Parlor," with Sloan's backing and Earl's drive the department grew in influence. In time, the renamed Styling Section expanded to 1,400 persons and became responsible for the styling of *all* GM models, with Earl eventually becoming a company vice president.

Styling at GM was influenced by two Sloan innovations. The marketing innovation involved the introduction of new models on an annual basis. New cars were unveiled at annual motor shows with great fanfare, a strategy that proved so effective that many car owners regularly traded in their cars every year or two. Sloan called it "dynamic obsolescence." His second innovation involved manufacturing. Previously, GM's car divisions had built their models independently, using their own distinctive "kit of parts." Sloan's idea was to make economies by having the divisions share interchangeable components. For example, an expensive Cadillac and a midprice Oldsmobile might use many of the same body parts and be distinguished chiefly by details, interior finishes, and the judicious use of chrome trim.

Sloan's approach relied heavily on the Styling Section to produce annual changes and to distinguish the different marques. Harley Earl pioneered the use of full-size clay models, although he was not trained as an artist and there is little evidence that he did much sculpting or sketching himself. A large, flamboyant man with a forceful personality, he was more like an impresario, setting general directions and critiquing the work of his staff. His philosophy was uncomplicated. "I have felt for a long time that Americans like a good-sized automobile as long as it is nicely proportioned and has a dynamic, go-ahead look," he wrote in a *Saturday Evening Post* article. He was an early proponent of consumer research. For example, when chrome began to appear as decorative trim on car bodies in the 1940s, he sent out ten of his staff, posing as newspaper reporters, to question car buyers in dealerships and on used car lots. The conclusion was that buyers, especially young buyers, liked chrome. So, Earl gave them chrome, lots of it. There is a story that the Styling Section had prepared a graphic presentation for him of two alternative chrome trim proposals for a particular car, but somehow there was a mix-up and the two alternatives were combined on a single car; much to the consternation of the stylists, Earl loved it and ordered them to proceed.

Earl is credited with the idea of the concept car. This was neither a preproduction prototype, nor a mock-up like the Briggs Dream Car, but an actual working automobile that was used to introduce new automotive ideas to the public. GM's first concept car, the Buick Y-Job, was designed by the Styling Section in 1938. The striking two-seater was displayed at motor shows and other public events, and later became Earl's personal car; he is said to have logged 50,000 miles by the time the car was retired in the 1950s. Some of the novel features, such as power windows and a convertible top that folded itself into a rear compartment, would later show up in production cars; others, such as power-operated concealed headlights and flush door handles, did not catch on

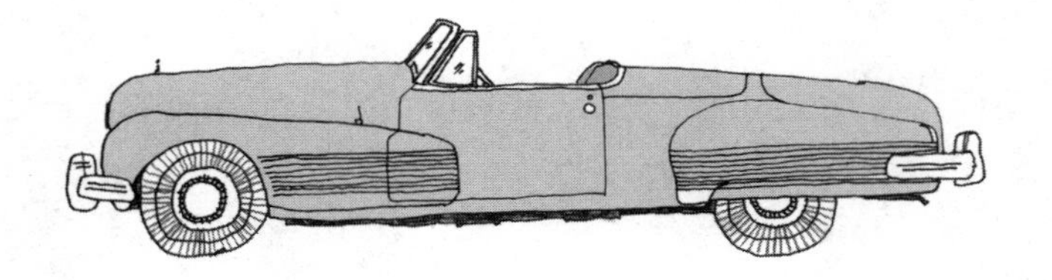

Buick Y-Job, 1938

immediately.* The eight-cylinder engine of the Y-Job produced 150 horsepower, twice as much as the LaSalle 303, but it was the body design that made the strongest impact on the public. "My primary purpose for twenty-eight years has been to lengthen and lower the American automobile, at times in reality and always at least in appearance," Earl observed. In the Y-Job, he achieved both goals. The roadster, built on a lengthened Buick chassis, measured more than seventeen feet long and was about ten inches lower than most contemporary sedans; the thirteen-inch wheels, rather than conventional sixteen-inch wheels, lowered the car still further. The massive fender skirts, the streamlined chrome accents, the wrap-around bumpers, and the sculpted body were in tune with the current Art Deco and Moderne styles that were affecting architecture, furniture, and graphics. The body design, which did away with separate running boards and fenders and emphasized an uninterrupted body, was referred to as pontoon styling. It is hard to believe that the Y-Job was a result of the same hand that had produced the slab-sided LaSalle only a dozen years earlier.

The Series 62 Cadillac, which arrived in 1940, showed the influence of the Buick Y-Job: a long, streamlined pontoon body and raked split windshield. The following year saw small modifications to the front grille and chrome trim. In 1942, the bodywork

* Concealed headlights were first used on the 1962 Lotus Elan, and flush door handles appeared even later, on the 2012 Tesla Model S and the 2017 Range Rover Velar.

Cadillac Series 62 Club Coupe, 1946

was dramatically redesigned so that the lengthened fenders overlapped the doors. Car production ceased during the war years and did not resume until 1946. The first postwar generation of Cadillacs was basically the delayed 1942 models, and the only radical design departure was the striking 1946 Club Coupe. The car was almost a foot longer than the Y-Job, and its sleek fastback rear recalled the prewar Tatra 77. Not that Earl was concerned with aerodynamics, as evidenced by the oversized fenders and the bulging front, with its massive chrome egg-crate grille.

SLOAN AND EARL HAD a radical effect on American car design. Previously, cars had been the purview of engineers, who designed the engine and chassis for stability and handling, the subsequent bodywork being governed chiefly by production efficiency. Now it was styling that set the pace: the body was designed first, and the mechanical components such as engines and transmissions were fitted in later. A car that typified this trend was the 1947 Studebaker, the work of Virgil Exner (1909–1973). Exner, born in Ann Arbor, Michigan, attended art classes at Notre Dame University and worked as an illustrator in an advertising agency in South Bend, Indiana. His illustrations of automobiles came to the attention of Earl, who hired the twenty-four-year-old for the nascent Art and Color Section. Exner fulfilled his promise, and in only two years was put in charge of the Pontiac design studio. During the Depression, Exner left GM and went to work for Raymond Loewy. The French-born Loewy (1893–1986) was a celebrated industrial designer—a field he had largely invented—known for creating corporate logos for Shell Oil and Lucky Strike cigarettes, and for styling a range of consumer products from refrigerators to office equipment, as well as industrial and transportation machinery—his clients included International Harvester, the Pennsylvania Railroad, and TWA.

The Studebaker Corporation, which dated from the first days of motoring, had recently emerged from receivership and had hired Loewy's firm as part of its comeback. The strategy produced a string of popular cars, and Exner headed up the Loewy auto design studio, which was located in South Bend, near the Studebaker plant.* Exner and Loewy did not get along, and in 1944, after Exner did some work for Studebaker on his own account, he was fired. A Studebaker vice president who was dissatisfied with Loewy's work hired Exner to do design work on the new Studebaker Starlight coupe. The coupe used a time-tested six-cylinder engine and was unremarkable mechanically, but Exner's body design was a radical departure from convention. The two-door five-seater resembled a roadster and had a smoothly sculpted body with a minimum of chrome. The trunk was roughly the same length as the hood. An unusual feature was the cockpit-like rear window that consisted of four curved glass panels. A wraparound rear window is of questionable practical value, but it caught the public's attention.

The Studebaker Starlight represented a further stage in the stylistic evolution of American car design. In art historical terms, the cars of the 1920s such as the Chrysler Six, the Ford Model A, and the LaSalle 303, with their straight sides and sharply defined verticals and horizontals, could be described as "classical." The following decade saw the billowing sculpted forms of the Lincoln-Zephyr, the Ford Model 68, and the Cadillac Club Coupe, opening a door to the baroque. In art historical terms, baroque is

* The main Loewy office, with more than a hundred staff, was in New York, and there were branches in Chicago, London, and Paris.

Studebaker Starlight coupe, 1947

succeeded by rococo, and Exner's light touch and delicate treatment of the rear window have some of the qualities of that effervescent theatrical style.

The success of the Starlight coupe created problems for Exner. Loewy was unhappy with Exner's involvement, and since Loewy was the public face of Studebaker, the company credited him with the design and felt obliged to let Exner go. By now he was well known in the industry, and he soon joined Chrysler as head of its Advanced Styling Group. Following the debacle of the Airflow, and the setback of Walter Chrysler's early retirement, the Chrysler Corporation was intent on returning to the forefront of car design. To facilitate this goal, the company formed a collaboration with Carrozzeria Ghia, a venerable Turin coachbuilder that had been founded by Giacinto Ghia (1887–1944) in 1916 and built specialized car bodies for Alfa Romeo, Lancia, and Fiat. An English-speaking Ghia executive, Luigi Segre (1919–1963), who a few years later would become the owner of the firm, developed a close working relationship with Exner. The result was the Chrysler K-310 concept car.

The K-310 used the chassis of the Saratoga, Chrysler's sporty luxury model, and added a 170-horsepower V8 engine, but the body, designed by Exner, was entirely built by Ghia in Italy. It was a large car, more than eighteen feet long, and although it sat six, the two-door coupe resembled a roadster. Exner's design included an egg-crate grille, exposed wire wheels, bulging fenders, flush door handles, and unusual gunsight taillights. The rear

Chrysler K-310, 1951

deck had a faux spare-tire bulge. The car was unveiled at the 1952 Paris Auto Show, and its combination of American flamboyance and Italian craftsmanship was well received. Although the K-310 and several later Ghia-built concept cars were widely praised, the financially strapped Chrysler was unable to put any of these adventurous models into production.

In 1953, the Museum of Modern Art in New York, which two years earlier had staged a popular show on cars of the 1930s and 1940s, organized an exhibition of postwar cars. *Ten Automobiles*, which was mounted in the museum's sculpture garden, featured seven European and three American models. The European cars were four roadsters—the French Ford Comète, the Simca 8, the MG TD, and the Porsche 1500—and three limited-production luxury cars: the Aston Martin DB2, the Lancia Gran Turismo, and the Siata Daina 1400. Two of the three American cars, the Cunningham C-4 and the Nash-Healey, were sports cars with bodies designed—and, in the case of the Cunningham, built—by Italian coachbuilders. However, the third was an all-American product: the Studebaker Commander Starliner.

The Commander Starliner was one of the most beautiful American cars of that decade. There were three models: a four-door and a two-door coupe, and a striking two-door hardtop. All models sat five. The Starliner was low for an American car, with a streamlined front and a similarly sloped trunk. The only chrome was on the bumpers, the undecorated hubcaps, and the minimal grille—just two narrow horizontal slots. Otherwise, the smoothly

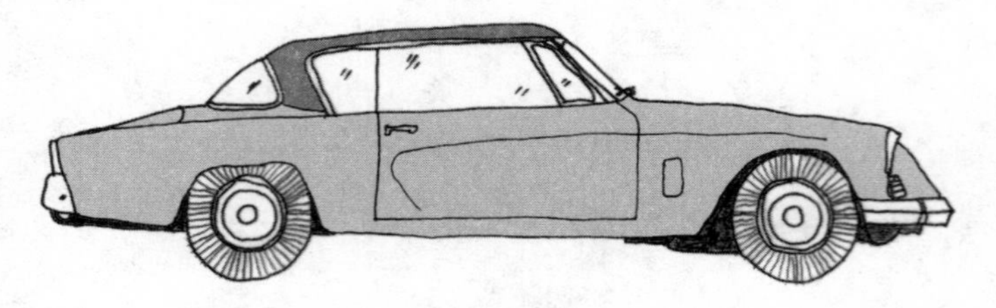

Studebaker Commander Starliner
hardtop coupe, 1953

sculpted body was devoid of trim. The Starliner was a sharp contrast to most American cars of the period—delicate rather than burly. The museum catalog described the Starliner as "the first American mass-produced car to adapt design characteristics of European automobiles." In fact, most of the European cars in the exhibition were rounded fastbacks, very different from the Studebaker; it was the Europeans who would adapt the knife-edge aesthetic of the Starliner, as would become evident when Citroën unveiled the DS two years later.

The design of the Starliner was the work of the Loewy automobile team, led by Robert E. Bourke (1916–1996), a Chicagoan who had been hired in 1941 by Exner. Although Studebaker promoted the Commander Starliner as the "Loewy Coupe," Raymond Loewy was not skilled in car design, and he gave Bourke a free hand. Bourke later recalled that the team started by modifying existing Studebaker designs, but was dissatisfied with the result. "I talked to [Loewy] on the phone one day, and I told him, 'I want to change it and change it completely, because I have thought it's not the way we should go.' And all he said was, 'Okay, Bob, go ahead, go ahead.' And with that, he went to Europe, and he was gone for quite a number of weeks, and when he came back, the car was done." Despite its striking appearance, the Starliner coupe was not a commercial success. Studebaker continued to face financial difficulties, and the resulting problems with delivery delays and quality control discouraged buyers. Moreover, although the graceful Starliner looked fast, with a time of seventeen seconds to go from zero to sixty miles per hour, it didn't deliver.

While Studebaker's entire production of cars and commercial vehicles had fallen to about 116,000 in 1955, Ford was building almost 1.5 million vehicles annually. One of its new models that year was a radical departure from convention: a two-seater. Ford had not built a roadster since 1932. The new car was the pet project of Franklin Q. Hershey (1907–1997), a Detroit native who had grown up in Southern California and worked for a Pasadena coachbuilder that designed custom bodies for Duesenbergs and Pierce-Arrows. Hershey joined General Motors, where he became a protégé of Harley Earl, and later head of the Special Car Design Studio. He was responsible for the 1948 Series 62 Cadillac, whose small tail fins—the first in the industry—were said to be inspired by the twin tail fins of the Lockheed P-38 Lightning fighter aircraft. Although the tail fins had no practical function, the postwar period saw the spread of air travel, so aeronautic references resonated with the public.

After leaving GM and working on his own and at Packard, Hershey became the chief stylist of the Ford division (Ford, Mercury, and Lincoln were separate divisions within the Ford Motor Company). He began the two-seater project in 1952 after learning that Chevrolet was about to launch a two-seater sports car—the Corvette (see chapter 8). At that time, almost all sports cars in the United States were British, German, or Italian imports, and it seemed to Hershey that there was an opportunity for another US-made model. He took it upon himself to develop a design, assisted by William Boyer (1926–2003), a young Pratt Institute graduate Hershey had recruited from GM. Styling at Ford was not an autonomous department—there was no Harley Earl paterfamilias. Rather, it was part of engineering, and when the engineering executives learned of Hershey's project, they were not supportive. The whole thing might have ended there, had Henry Ford II not decided that the company should have a sports car in its lineup.* Hershey, who had already built a full-size clay model, was given the go-ahead.

Hershey had to work under several constraints. Imported European sports cars were quirky machines, intended for enthusiasts and produced in small numbers—in 1952, Porsche's total American sales amounted to fewer than three hundred cars. A Ford, which would be produced in much larger numbers, had to be comfortable and practical as well as sporty. Thus, the new car was given power seats, a telescoping steering wheel, and optional power brakes and power steering. The transmission was a three-speed manual, with optional automatic transmission and power windows. The convertible—the only model—came with a removable fiberglass top, and an optional fabric top that folded into the trunk. The car used standard Ford parts, partly to minimize tooling costs and partly to ensure ease of maintenance, so that the car could be serviced by any Ford dealer.

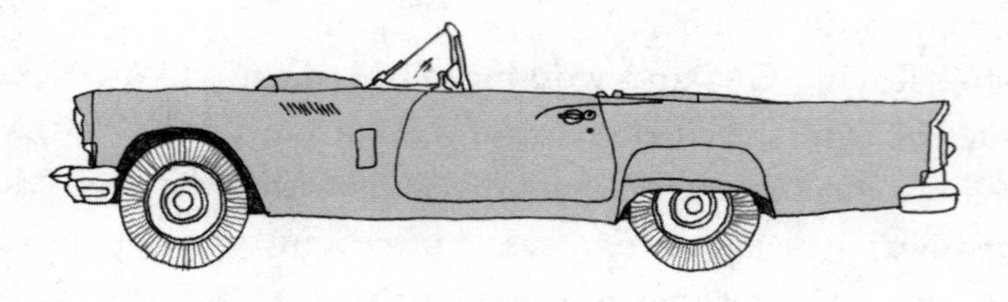

Ford Thunderbird, 1955

The Ford two-seater, which was christened Thunderbird, did not have the streamlined curves of an Aston Martin or a Porsche; instead, the body had the knife-edge crispness of the Studebaker Commander. Like the Commander, the hood sloped below the

* Henry Ford II was a motoring enthusiast who owned two Cisitalia 202 sports cars, 1947 classics of which only 170 were built. He later supported the successful development of the Ford GT40 racing car that beat Ferrari at Le Mans.

fender line; so did the trunk, which created a modest fin-like effect. Hershey resisted all attempts to add the chrome that was characteristic of other Ford models at that time. "It was a clean, simple design that did not need accent trim," he said in a later interview. An understated cowl hood scoop hinted at a racing heritage, although it was actually required to create extra clearance for the large engine. The 193-horsepower Ford Y-block, which had replaced the old flathead V8 the previous year, had a four-barrel carburetor and produced a top speed of 118 miles per hour. Nevertheless, Ford played down the car's sportiness and advertised the Thunderbird as "a distinguished kind of personal car that combines high performance and high style for a whole new world of driving fun." But as the *Motor Trend* review put it, "Although the Ford Motor Company is the first one to deny it, they have a sports car in the Thunderbird, and it's a good one." This judgment was borne out by the many T-Birds that competed in rallies and at Sebring, Daytona, and other racetracks.

The handsome Thunderbird was a publicity success. Marilyn Monroe and Arthur Miller bought one; so did Frank Sinatra. "There's a touch of Thunderbird in every Ford," boasted an ad. Although 16,000 Thunderbirds were sold in the first year, Fords generally sold in the hundreds of thousands, and many in the company saw no future for a specialized roadster. Robert McNamara, one of the financial "Whiz Kids" hired by Henry Ford II and head of the Ford division, felt that the Thunderbird brand had proven itself, but that it needed to move into the mainstream. That meant being transformed into a four-seater. Hershey argued vainly for at least keeping the original two-seater in production. The redesigned 1958 Thunderbird was a bulky and overdecorated four-seater with none of its earlier sporting grace, but sales doubled, and with a price of $1,000 more than the two-seater, the car became a profitable product. "As a business decision, it was well thought out; as far as Thunderbird buffs go, it was a disaster," observed William Boyer, who was put in charge of styling the new model after Franklin Hershey resigned. Hershey would go on to have a long career as an industrial designer, but henceforth stayed out of the car business.

I drove a Thunderbird once. It was 1963, I was a college student, and the car belonged to my girlfriend's father, a wealthy business executive. I was taking Liz to her prom at a private girls school, and her father casually handed me the keys. "Why don't you take my car?" I was shocked. I did not own a car, and every outing in my father's Vauxhall was preceded—and often followed—by a cautionary lecture. I don't remember much about the T-Bird—it was a third-generation four-seater, black with a leather interior, a large car compared to the Vauxhall. Mostly, I was nervous about scratching it, but I managed to return the car unscathed.

IN 1957, THE LAST year of the two-seater Thunderbird, Chrysler unveiled a group of new models that shook up the car industry. Instead of the small, incremental changes that characterized GM's dynamic obsolescence strategy, Virgil Exner, who was now in charge of styling at Chrysler, pulled out all the stops and produced a strikingly new series of designs that were marketed as the Forward Look.

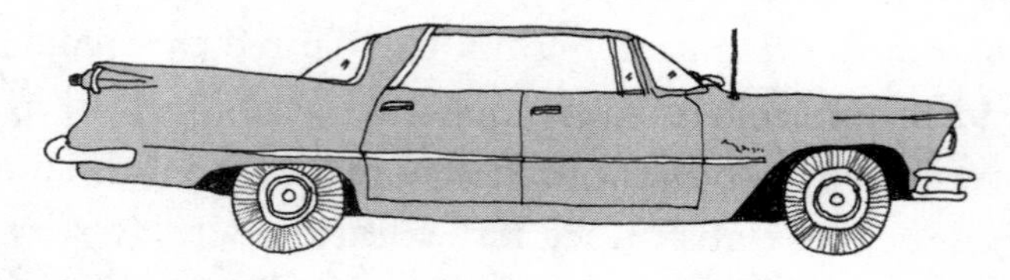

Imperial Crown, hardtop coupe, 1957

The Imperial Crown, Chrysler's top-of-the-line model, typified the Forward Look. The large car was almost nineteen feet long, with a width of eighty-one inches—almost four-abreast seating. The massive 5.4-liter V8 engine, the largest in a production car at that time, was paired with a push-button automatic transmission and produced 350 horsepower, driving the nearly 5,000-pound car at a top speed of 120 miles per hour. Imperials had had power steering since 1951, and air-conditioning since 1953. The windshield and the rear window were curved—an industry first—and the body included quad headlights and an unusual overlapping roof detail. The sloping rear deck had the faux spare-tire bulge that had become an Exner signature. The most striking body features of the Imperial Crown were the prominent tail fins—not Hershey's modest version but aircraft-size fins with integrated rear lights that resembled gunsights. The high-quality Imperial was produced on its own assembly line. At more than $5,000, the four-door hardtop was expensive, and it looked expensive. The Hollywood actress Lauren Bacall owned one in Horizon Blue.

Cadillac Series 62
hardtop coupe, 1959

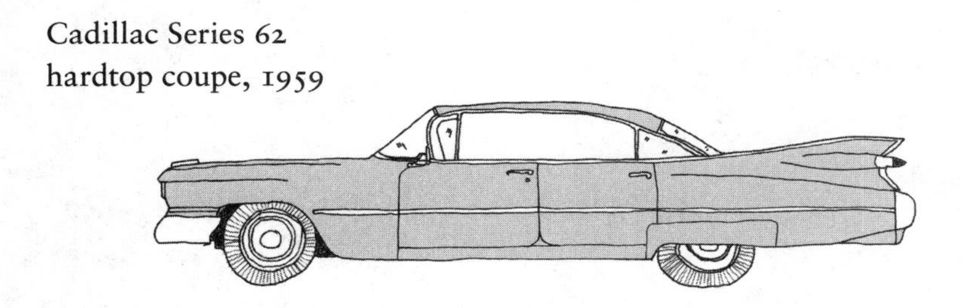

Chrysler's Forward Look caught Harley Earl and his team by surprise. The sleek Imperials made the bulging GM cars look downright frumpish. It was too late to do anything about the 1958 models (retooling required a lead time of at least two years), but Earl, who was set to retire the next year, had the 1959 lineup scrapped and ordered a crash redesign program. This work was largely in the hands of his groomed successor, Bill Mitchell, whom Earl had recruited in 1935, and who had been responsible for the 1955 Chevrolet Bel Air. Mitchell continued his mentor's approach in the 1959 Cadillac de Ville, which was long and low. Whereas most cars gave pride of place to the front grille, the Cadillac's rear was, if anything, more striking than the front. A rear grille mimicked the front; the huge fins, trimmed in chrome, had twin bullet-shaped taillights that were meant to resemble a fighter jet's thrusting afterburner; and the conelike nacelles actually doubled as exhausts in some models.

Cars like the 1959 Cadillac represented the climax of what is sometimes called the Tail Fin Era. The biggest fins were on American models, but modified fins appeared on European cars—Volvo, Saab, Auto Union, Mercedes-Benz, even the Soviet Moskvitch. Fins appealed to designers because they could be large or small, histrionic or restrained. The following year, as the American economy entered a two-year recession, Mitchell, now in charge of styling, scaled back many of Earl's most exuberant enthusiasms, including fins and chrome trim, not to mention afterburner-like nacelles.

CHAPTER
SIX

ALSO-RANS

At a quick glance, the car resembles a Cadillac that started smoking too young.

—TOM MCCAHILL
ON THE HENRY J

THERE ARE SIMILARITIES BETWEEN DESIGNING CARS AND designing buildings. Both require balancing technical demands and human needs, and both deal with aesthetics and taste. There is a major difference, however; the fate of a building rarely depends on its users—once built, it's there for better or worse, and it's not going away. But as we have seen with the Chrysler Airflow, if buyers don't warm to a car, it can go away very quickly. This happened to a number of new American models during the postwar period.

During the Second World War, the US government required the major automobile manufacturers to stop producing civilian vehicles and switch to making military transport, armaments, and aircraft. When peace was declared, there was pent-up demand for cars, but the Big Three needed time to develop new models, which gave newcomers an opportunity to enter the field. One of these parvenus was the grandly named Tucker Corporation. Preston T. Tucker (1903–1956) had an unusual background for a car maker; he was neither a machinist, like Walter Chrysler; an engineer, like Alfred Sloan; nor an MBA, like Robert McNamara. Tucker was a promoter. He grew up outside Detroit, and after dropping out of high school he pursued a boyhood interest in automobiles, working for Cadillac as an office boy, manning a Ford assembly line, running a gas station, and taking a series of sales jobs in car dealerships. He briefly became involved with racing in Indianapolis. Despite lacking a technical background, during wartime he started a small engineering company that developed an armored car, a fighter plane, and a gun turret, although few of these projects went beyond the prototype stage.

After the war, Tucker returned to his automotive interest and set out to become a manufacturer. He hired an illustrator to draw a futuristic-looking car that featured in a full-page advertisement in *Mechanix Illustrated* and *Science Illustrated*: "More like a Buck Rogers Special than the automobiles we know today, the Tucker Torpedo . . . will incorporate a series of spectacular engineering innovations that conservative auto manufacturers have classified as 'at least five or six years off.'" An accompanying photograph showed what looked like a prototype but was actually a quarter-scale model. The following year, an admiring article on the Torpedo in *Popular Mechanics* elaborated Tucker's extravagant promises: "This revolutionary car carries six passengers and will be in the 'medium price class'—$1,500 to $1,800." Tucker shied away from taking bank loans and instead financed his project by selling dealer franchises for the future car. His plan for an early public stock offering was squelched by the Securities and Exchange Commission, which required him to first build a working prototype. The enterprising Tucker formed a board of directors, assembled a team of engineers, and leased a converted Dodge plant in Chicago that had been used during wartime to build aircraft engines for B-29 bombers.

Tucker was full of ideas for his "Car of Tomorrow." The most radical was an unusual engine with overhead valves operated by oil pressure rather than a camshaft, connected to a hydraulic drive and torque converters on each driving wheel. His fledgling company did not have the resources—or the time—to develop such a radical innovation, and after many failed attempts he made do with a converted six-cylinder water-cooled aircraft engine and a conventional transmission. Similarly ambitious plans for improved brakes, lightweight magnesium wheels, four-wheel independent suspension, and fuel injection fell by the wayside. One of Tucker's ideas that did survive was the rear location of the engine, a first for an American production car. Also unusual was his concern for safety. The car included a protective structural cage and a roll bar, as well as a front crumple zone, a collapsible steering column, a shatterproof pop-out windshield, and a padded dash. A distinctly quixotic safety feature was a padded cavity beneath the dashboard. "Front seat occupants can drop into this space in a split second in case of unavoidable collision," according to a Tucker advertisement.

The body was largely the work of Alexander S. Tremulis (1914–1991). Tremulis was a self-taught automotive designer who, at only twenty-two, had been head stylist for the Auburn-Cord-Duesenberg Company. He was working for a Chicago industrial design firm when Tucker hired him to turn the early promotional sketches of the Tucker Torpedo into an actual car. The entire development process was squeezed into less than two months, and a month later, in June 1947, Tucker unveiled a working prototype. The renamed Tucker 48, a four-door five-seater, was a combination of new and old. The low, streamlined body included unusual doors that extended into the roof, and an eye-catching cyclops headlight that swiveled as the car turned. On the other hand, although the Chrysler Airflow had introduced a one-piece windshield two decades earlier, the Tucker had an old-fashioned split windshield, and the bulging fenders and sloping fastback recalled Harley Earl's Cadillac Club Coupe of the previous year. Indeed, the Tucker 48 was comparable in size, weight, and power to the Club Coupe. And in price. The Tucker 48 cost $2,450, slightly more than a Cadillac and hardly in the "medium price class."

Tucker established a small assembly line and started building cars, but henceforth, things did not go well. Cash flow was a con-

Tucker 48, 1948

stant problem. Despite having raised $6 million from the sale of dealer franchises, and $15 million from a stock offering, Tucker was in arrears to the federal War Assets Administration, which owned the Chicago factory. The $2 million he raised from preselling accessories to future car buyers proved to be his undoing. The Justice Department judged this unorthodox fundraising method to be fraudulent, and took Tucker and his board to court on the basis that the projected car was a hoax. A grand jury ultimately found the defendants not guilty, but by then it was too late; the factory had been confiscated for nonpayment of rent and the Tucker Corporation was in bankruptcy. The fifty-one cars that had been built were sold off with the other assets. A subsequent plan to develop a Brazilian sports car came to naught, and Tucker died of lung cancer, only fifty-three.

UNLIKE PRESTON TUCKER, Henry J. Kaiser (1882–1967) was an experienced industrialist. His construction company had been part of a consortium responsible for building the Hoover Dam, and during the war the Kaiser Shipyards in Seattle had successfully launched aircraft carriers, troop transports, and cargo vessels, the so-called Liberty Ships.* In 1945, in partnership with experienced car builder Joseph W. Frazer (1892–1971), Kaiser formed the Kaiser-Frazer Corporation, which acquired the assets

* The managed-care giant Kaiser Permanente was cofounded by Kaiser initially to provide health insurance and care for the Hoover Dam construction workers, and later the workers of the Kaiser Shipyards.

of the Graham-Paige car company that Frazer had headed before the war.

Kaiser-Frazer manufactured a variety of full-size cars, but one of Henry Kaiser's pet projects was a new model that was launched in 1950. Thanks to his well-earned reputation, he secured a federal government loan of $44 million (more than half a billion today), undertaking to build an affordable sedan that would seat five, have a cruising speed of fifty miles per hour, and cost no more than $1,300. The resulting two-door coupe was an unremarkable design with a complicated chrome grille and a suggestion of tail fins. Advertisements often showed children next to the car to make it appear larger; with a wheelbase of only 100 inches, the Henry J was small—a compact car before there were compact cars.

Kaiser-Frazer
Henry J, 1950

The Henry J was a conflicted concept—a budget-minded compact car with full-size pretensions. To reduce cost, the trunk was accessed by folding down the rear seat back—there was no openable lid. In addition, the rear windows did not open, and the spartan interior lacked a glove compartment, armrests, and fresh-air ventilation. The four-cylinder engine, bought from Willys-Overland, was the same one used in the Jeep CJ and put out only sixty-eight horsepower. The small, light car could get thirty-five miles per gallon, although that was hardly a major selling point at a time when gasoline cost twenty-seven cents a gallon. The Henry

J was not a commercial success, and much of the first year's production of 82,000 remained unsold. The problem was that most budget-minded buyers preferred to buy a used full-size car rather than a new compact, particularly if the latter was widely seen as "cheap." Kaiser-Frazer eventually added an openable trunk and an optional six-cylinder engine, and made an arrangement with the Sears, Roebuck catalog company to sell the car as a rebadged "Allstate." But to no avail. The second year's production fell to 30,000. In the third year, after prices were slashed, only 17,000 cars were produced, and Henry Kaiser finally pulled the plug on the project.

UNLIKE THE TUCKER CORPORATION and Kaiser-Frazer, Nash Motors was an established marque. The company was founded in 1916 by Charles W. Nash (1864–1948), who had been the successful fifth president of General Motors. The Nash company became known for well-made, reliable, and affordable cars that found a ready market even during the Depression. Nash retired in 1932, and his handpicked successor was George W. Mason (1896–1954). Mason, an engineer, was not a car man—he had turned Kelvinator into the second-largest refrigerator maker in the country. To secure Mason's services, Nash had to agree to acquire the company, which was not as unusual as it sounds (General Motors owned Frigidaire) and Kelvinator later made air-conditioning compressors for Nash automobiles.

Nash-Kelvinator's vice president of engineering was a Finnish-born Swede, Nils Erik Wahlberg (1886–1977). He had studied engineering in Zurich, and after emigrating to the United States he worked for General Motors, where he met Charles Nash, who recruited him for his new company. The talented Wahlberg was responsible for many technical refinements in Nash autos, including America's first unibody construction.

In 1949, Nash introduced two new postwar models, the Ambassador and the 600. During the war, Nash-Kelvinator had

Nash 600, 1949

produced aircraft and aircraft engines. Wahlberg, with access to a wind tunnel, grew convinced of the benefits of aerodynamic design, and the new cars were the first all-out attempts at streamlining since the Chrysler Airflow debacle fifteen years earlier. The lead seller was the 600, which was shorter than the Ambassador, with a 112-inch wheelbase and a smaller engine, a flathead six-cylinder power plant that produced eighty-two horsepower and powered the lightweight car at a top speed of eighty-five miles per hour. The "600" referred to the fact that the car could travel six hundred miles on a tankful of gas. Nash marketed the streamlined styling as "Airflyte," and the car is estimated to have had a drag coefficient of 0.52. Like most aerodynamically designed cars, the 600 was a fastback, and the striking unibody was unlike any car of that period—effectively, an American Tatra. In many ways, it is the Nash 600—not the Tucker—that qualifies as a "Car of the Future." There were no fenders, and the swelling sides of the car were smooth, with a single chrome accent strip; the windshield was one piece of curved glass. The wheels were shrouded, which gave rise to the "Bathtub" nickname. The modest grille, on the other hand, was conventional; Nash wasn't about to make the same mistake as the Chrysler Airflow.

The Nash 600 was a comfortable family car. Seating five, it came in two-door and four-door models. The "Weather-Eye" heating system used a fan to maintain a positive pressure inside the car, and recirculated filtered fresh and heated air, depending on the season, which was a significant innovation at a time when most car heaters were fairly crude. Nash was the first American manufacturer to offer seat belts, which were a factory installed option on all its 1949 cars.* The interior of the car was styled by Helene Rother (1908–1999), a German-born art school graduate whose Detroit studio was responsible for many Nash models. The well-equipped interior, which included a clock and an AM radio, was as thoughtfully designed as the mechanical systems. The speedometer and other instruments were grouped in a pod that was mounted on the steering column immediately behind the wheel. The glove compartment was replaced by a convenient storage drawer. The fully reclinable front seats allowed an unusual option: a twin bed. This was a common feature on Swedish cars at that time and was likely Wahlberg's suggestion.

Hudson Hornet, 1951

Two years later, the Nash 600 had a competitor, the Hudson Hornet. Founded in 1909, the Hudson Motor Company had grown rapidly, and by the 1930s it was the third-largest car company, after Ford and Chevrolet. Although Hudson had since slipped from that position, the company continued to have a reputation for mechanical innovation. The Hornet had an unusually powerful six-cylinder engine that put out 145 horsepower and a top speed of ninety-three miles per hour. The full-size six-passenger car incorporated a Hudson innovation, a "step-down"

* The Nash seat belts were not popular and encountered resistance from car buyers, many of whom requested them to be removed.

body whose floor was inside—instead of on top of—the perimeter chassis frame, which lowered the center of gravity of the car and resulted in exceptional handling. The Hornet was three inches lower than the Nash 600, and had rear-wheel skirts and a streamlined body with flashy chrome accents. While not quite as sleek as the 600—the Hornet had a drag coefficient of 0.60—it was a fast car. A version of the engine with twin carburetors and a higher compression ratio was an option in 1952 and increased the output to 170 horsepower, making the Hornet the fastest production car on the market, and a Hornet factory team dominated national stock car racing for years.

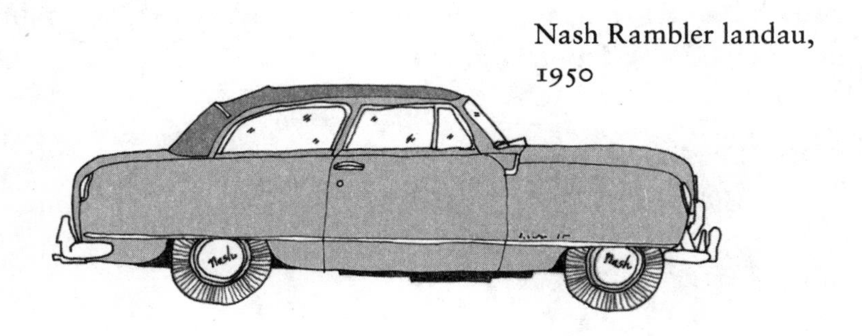

Nash Rambler landau, 1950

Despite their advanced design, the Nash 600 and the Hudson Hornet had difficulty competing with the new models being introduced by the Big Three. Nash did not have the financial resources to develop a V8 engine, which had become the new standard for American full-size sedans, and decided to turn its attention to a smaller car. In 1950, it introduced the Rambler, a compact version of the Airflyte streamlining concept. The two-door five-seater had a 100-inch wheelbase, but it was not marketed as a budget car; instead, the first Rambler was an upscale convertible that Nash called a "landau." The unibody construction required that it be a fixed-profile convertible—that is, the rigid body sides supported a center fabric roof that unfolded. This was similar to the Fiat Topolino and the Citroën 2CV, except that the accordion roof was electrically operated. At this time, Nash had a relationship with Carrozzeria Pinin Farina, and it is likely that the Italian coachbuilder was responsible for some of the body details. There was nothing cheap-looking about the fully equipped car, which had a smart Rother-designed interior, used the same six-cylinder engine at the Nash 600, and sold for $1,800. The following year, having established the Rambler as a desirable marque, Nash added a sedan, a hardtop, and a station wagon. In 1954, air-conditioning became an option on several Nash models, including the Rambler. Nash was not the first manufacturer to offer this feature. Packard had a short-lived venture into air-conditioning in 1940, and in 1953 Chrysler and General Motors included expensive systems on their high-end models. But the compressor developed by Kelvinator was the first compact front-end system that combined cooling, heating, and ventilation at an affordable price.

Nash Metropolitan convertible, 1953

The success of the Rambler encouraged Nash to develop an even smaller car, the Metropolitan, which was a radical departure from convention in several ways. To begin with, it was very small—today, it would be called a subcompact—with an eighty-five-inch wheelbase that was ten inches shorter than the Volkswagen Beetle, which was then becoming available in the United States. The Metropolitan was powered by a 1.2-liter four-cylinder engine that put out forty-two horsepower and reached a top speed of seventy-five miles per hour; fuel consumption was forty miles per gallon. The two-door car, which came as a convertible or hardtop, had two front seats and a rear bench suitable for "small children, pets, or packages," according to Nash. The trunk was accessed by lowering the lockable rear seat back—there was no opening trunk lid (one was eventually added). The spare wheel, mounted on the exterior, was what was then called a "continental tire."

The unibody body was designed by William J. Flajole (1915–1999), a freelance stylist who had worked for Chrysler and General Motors before opening his own Detroit studio. Flajole, taking his cues from the Rambler, gave the Metropolitan rounded Airflyte styling, a simplified grille, and skirts over the wheels. The notched "pillow-style" doors were an unusual feature. The interior included leather-trimmed upholstery, Nash's Weather-Eye heating system, a basic AM radio, a cigarette lighter, and dual sun visors. Although Nash's message was "luxury in miniature," the design included such discreet economies as no quarter-glass vent windows, interchangeable left and right doors, a lidless trunk, and a doorless glove compartment. Many of the advertisements featured women drivers, because the Metropolitan was specifically marketed to suburban women as a second family car, useful for dropping the children off at school and for shopping errands. Nash promoted the Metropolitan as "Smart Foreign Car Flavor . . . American Car Comfort and Convenience." Despite being designed in the United States, this really was a foreign car. The bodies were made in Birmingham by Fisher & Ludlow, a leading British body maker, and finished by Austin using its own mechanical components, engines, gearboxes, rear axles, suspensions, and brakes. The saving in tooling costs—and a strong American dollar—made this arrangement economically advantageous to Nash.

In 1953, the year that the Metropolitan was introduced, Hudson unveiled the compact Jet, the company's answer to the Rambler. But by then, the novelty of smaller cars was wearing off; in addition, a price war between Ford and Chevrolet reduced the prices of entry-level full-size cars. Hudson, which had overinvested in tooling for the Jet, ran into trouble, and the following year the company was obliged to merge with Nash, to form the American Motors Corporation. AMC discontinued the Jet and the Hornet, and two years later halted production of the Rambler (the marque was revived in 1958).

Surprisingly, the little Metropolitan outlasted them all, remaining in production until 1962. By then, almost 90,000 had been sold in North America, a paltry number at a time when the entry-level Chevrolet Bel Air sold more than five times as many *in a single year.* Of course, the little Metropolitan was ahead of its time; had it appeared in 1973, when demand for compact cars exploded, it might have been a different story. But in the 1950s, American cars were getting larger, with greater-displacement eight-cylinder engines, gas was cheap, and the Metropolitan appeared out of step, effectively a two-seater that was neither very fast nor very sporty. Yet when *Mechanix Illustrated*, which ran a regular feature on users' experiences with their cars, sent its standard questionnaire to Metropolitan owners, it reported an unexpectedly enthusiastic response. "Never have we received such a large percentage of returns on the questionnaires. Never have we had so many comments written on the blanks. In other words, the Metropolitan owners know their cars and they love this tiny import bearing the Nash nameplate." Like the Volkswagen, the "Met" introduced Americans to the practical advantages and driving pleasures of small cars. And, like the VW, the Nash had character, which appealed to a certain kind of individual. Steve Jobs's first car, bought when he was fifteen, was a used Metropolitan.

THE ALSO-RANS DISCUSSED ABOVE were the work of either brand-new or smallish car makers, but even the Big Three could sometimes trip and stumble. The Ford Motor Company, despite having surpassed faltering Chrysler in sales, was still striving to catch up to General Motors. Although Ford's entry-level cars were selling briskly, the company was not well represented in the more profitable midprice sector. To rectify this situation, in 1954 Ford launched an entirely new division in the midrange category. To be competitive, the new cars, awkwardly named Edsel after Henry Ford's son, required a unique identity. That meant the Edsel needed to be "readily recognizable in styling theme from the nineteen other makes of cars on the road at the time," according to Roy Abbott Brown Jr., the Ford stylist in charge of the project.

The Canadian-born Brown (1916–2013) had grown up in Detroit, where his father worked as an engineer for Chrysler. After graduating from the Detroit Art Academy in 1937, Brown went to work for General Motors under Bill Mitchell, and after completing his military service, in 1953 he joined Ford. Brown's task with the Edsel was challenging because Ford had decided, for reasons of economy, to adopt GM's interchangeability strategy: the new car was to share its platform, body shell, and many mechanical components with existing Fords and Mercurys. Brown had worked on the Lincoln Futura, a concept car that became the inspiration for the Batmobile in the 1960s *Batman* TV series, and he brought a graphic, sci-fi sensibility to the Edsel. For example, he made the front grille vertical instead of horizontal, gave the body prominent concave scallops and a two-tone paint scheme, and designed unconventional gull-wing rear taillights. The Edsel definitely didn't look like other cars.

In the past, automobile styling had been guided by manufacturing economies, aerodynamics, or performance. The design of the Edsel seems to have been curiously arbitrary in that regard, more like difference for difference's sake.* This iconoclasm continued on the interior: the speedometer was a rotating dome instead of a dial, and the push-button automatic transmission was located in the center of the steering wheel, replacing the horn, which was activated by a horn ring. At the last minute, the headlights were changed to quads simply because this feature had appeared in many 1957 General Motors and Chrysler models. Other inno-

* Brown's design was never test-marketed, although Ford later did extensive motivational research among potential buyers of the Edsel as part of its marketing campaign.

Ford Edsel
Ranger, 1958

vative features were more sensible: warning lights, a remote-operated trunk, seat belts, and childproof rear door locks.

The Edsel was not simply a new car but an entirely new line of cars: no fewer than seven distinct models. The base-trim Ranger, and its upscale twin, Pacer, both used the Ford Fairlane platform; the Corsair and the top-of-the-line Citation were built on a larger Mercury platform; and the three station wagons included the entry-level two-door Roundup, the Villager, and the upmarket Bermuda, a fully equipped woodie. The sedans came in four configurations—two- and four-door coupes, and two- and four-door hardtops—and the Pacer and Citation were also available as convertibles. The Villager and Bermuda station wagons had six-passenger and nine-passenger versions. That added up to no fewer than twenty-three body styles! The Edsels were midprice, but they were not midsize; the Citation, for example, was two inches longer than a 1958 Cadillac hardtop sedan, weighed almost as much, and had a slightly more powerful engine.

The Edsel project cost Ford $250 million (about $2.7 billion today): $150 million to upgrade the Ford and Mercury plants, $50 million for new tooling, and $50 million for marketing. The 1958 launch kicked off with an hour-long Sunday night TV special—*The Edsel Show*—starring Bing Crosby and featuring Frank Sinatra, Louis Armstrong, and Bob Hope. The Emmy-winning show was a hit; the Edsel less so. Ford planned on selling 200,000 cars a year, but as sales reports came in from the 1,200 newly established Edsel dealerships, it quickly became apparent that this goal would not be met. The first year saw 64,000 sales. For the second year, Ford dropped the Corsair, the Citation, and two of the three wagons, replaced some of the odd features, such as the rotating-dome speedometer and the steering-wheel transmission (which confused many drivers), and toned down the body styling. But annual sales continued to fall—to 45,000. A harebrained scheme to stimulate sales by raffling off children's ponies at Edsel dealerships—"Bring your parents to an Edsel dealer," went the television ads—fell flat. Robert S. McNamara, who was now in charge of Ford, canceled the Edsel advertising budget, and in November 1959 the Ford Motor Company announced that it was discontinuing the entire program.

What went wrong? Almost everything. The car launch coincided with a yearlong economic recession, which drastically reduced sales of new cars, especially midprice ones. The recession also altered the kind of car that budget-minded middle-income buyers wanted—not an oversize vehicle with a gas-guzzling engine, but something smaller, more budget-minded.* There was also a problem with reliability. To save money, instead of building a new Edsel factory, Ford had slotted the car into its Ford and Mercury plants, creating confusion among the assembly-line workers and leading to poor quality control—cars were sometimes rushed to dealers with uninstalled parts in the trunk. The decision to establish a brand-new network of dealers was likewise questionable.

Whether the unorthodox styling of the car contributed to its demise has been debated. The vertical grille came in for ridicule—"an Oldsmobile sucking a lemon"—but there had been comparable vertical grilles on several successful high-end cars during the 1930s, notably LaSalles, Bugattis, and Alfa Romeos. What is not disputable is that the Edsel's was a flawed design; the front,

* The only car that posted an increase in sales in 1958 was AMC's inexpensive compact, the Rambler.

back, and sides didn't add up to an integrated whole, and instead appeared to have come from three different cars. Compared to Bourke's sophisticated Studebaker, Hershey's lean Thunderbird, and even Exner's extravagant Imperial Crown, Brown's gaudy Edsel lacked conviction.

THE 1960 RECESSION, AND the success of cars such as the Rambler and the Metropolitan, convinced the Big Three to enter the compact-car field. Ford and Chrysler took a conservative approach and introduced cars that were basically scaled-down versions of their full-size models, but General Motors was more adventurous. The head of Chevrolet was Edward N. Cole (1909–1977), who had been the chief engineer of the division and was responsible for developing the small-block V8 engine that had become the GM standard. Cole had designed rear-engine vehicles for the military during the war, and he championed the revolutionary—for America—proposal to make the new compact Chevrolet, called Corvair, rear-engined. The car was given an all-new air-cooled six-cylinder engine with aluminum heads and block. Other technical innovations included unibody construction, four-wheel independent suspension, and a rear transaxle that combined transmission, axle, and differential in a single unit. With a wheelbase of 108 inches, the Corvair was longer than the Rambler, but it was the smallest car in the Chevrolet lineup, and at less than 2,500 pounds a lightweight compared to full-size cars weighing in excess of four thousand pounds.

The body design of the Corvair is credited to Ned F. Nickles (1919–1987), who was head of Buick's design studio, working under Bill Mitchell, who had taken Harley Earl's place as head of the Styling Section. Although Nickles was a protégé of Earl, and was responsible for the flamboyant 1953 Buick Skylark convertible, he was also a sports car enthusiast who owned a Siata Daina, the sleek Italian two-seater that had featured in the Museum of

Modern Art car show. Nickles gave the Corvair a similarly unadorned form, with a wraparound beltline and almost no chrome, no spears or accents, no hood ornament, no grille, and definitely no fins. A low driving position reduced the car's height to fifty-one inches, which resulted in a particularly well-proportioned body. What had started as a budget model had emerged as an elegant and sporty coupe. David E. Davis Jr. of *Car and Driver* called it "the most beautiful car to appear in this country since before World War II."

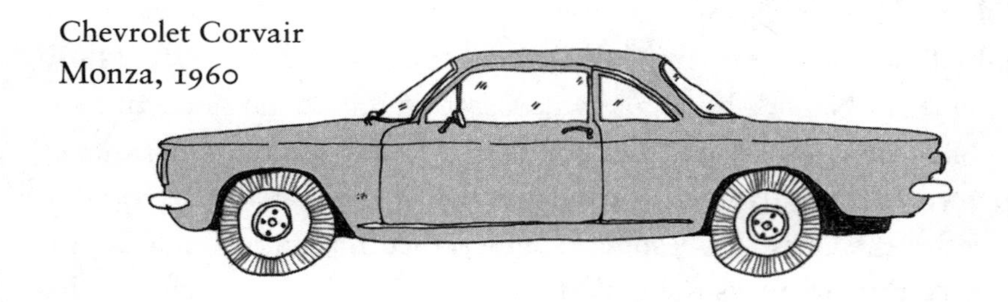

Chevrolet Corvair
Monza, 1960

Young buyers in particular were attracted to the Corvair, not only by its $2,000 sticker price but also by its styling and performance; the 2.3-liter engine produced eighty horsepower and a top speed of ninety miles per hour. A particularly popular model was the two-door Monza coupe, which had bucket seats, automatic transmission, deluxe trim, and special upholstery, and accounted for more than half of Corvair sales. The Monza taught an important lesson: buyers of compact cars were attracted to smaller cars, but they valued the same features as buyers of larger models.

By 1965, General Motors had sold 1.6 million Corvairs. That same year, Ralph Nader, a consumer activist, published *Unsafe at Any Speed*, a critique of the safety record of the American car industry. The best-selling book became a cause célèbre, and led to congressional hearings and the resulting National Traffic and Motor Vehicle Safety Act. Nader's chapter on the Corvair accused

GM of jeopardizing safety to increase profits, and denounced the car as inherently dangerous.* The Corvair did have a tendency to oversteer—that is, to turn more than the amount commanded by the driver. Oversteer is the result of a center of gravity that is more to the back of the vehicle, which was the case with the Corvair, with its heavy rear engine. Chevrolet instructed owners to keep high pressure in the rear tires and low in the front, a common solution in European rear-engine cars such as the Volkswagen Beetle and the Renault Dauphine. Many American owners ignored this directive, and, unused to rear-drive handling, experienced a number of accidents. By the time Nader's book came out, Chevrolet had installed front anti-roll bars that resolved the oversteer problem, but Nader's book had a devastating impact on sales. In 1969, Chevrolet, which had developed the sporty Camaro to compete with the popular new Ford Mustang (see chapter 8), decided to retire the Corvair. Despite its undeserved reputation, the Corvair was in many ways the most technically ambitious—and the most attractively designed—of the also-rans.

* A 1972 report of the federal National Highway Traffic Safety Administration concluded that Nader's claims that the Corvair was unsafe had no basis in fact.

CHAPTER
SEVEN

AUSTERITY

It'll still be fashionable when I'm dead and gone.

—ALEC ISSIGONIS ON THE MINI

AUTOMOBILE MANUFACTURING IN POSTWAR EUROPE WAS very different than in the United States, and the difference had a major impact on car design. Gaston Fleischel, a French automotive engineer and inventor of a pioneering automatic transmission introduced by Peugeot in the 1930s, summed it up: "While American manufacturers are leading far ahead with roomy, luxurious, silent, and very comfortable cars, and achieving higher mass production, the European engineers are confronted with customers impoverished by two world wars, sometimes both fought on their soil, with heavy taxes on gasoline and engine displacements, not to mention the tax collector looking for large cars as a sign of wealth for higher taxation.* They are induced, therefore, to produce smaller cars, and are more concerned with safety, low cost of performance, and general operation." European conditions produced cars such as the Fiat Cinquecento in Italy and the Citroën 2CV in France, which were basic, low-budget vehicles that American critics sometimes referred to as "shoehorn cars," because of their confined interiors.

* "Sign of wealth" was a reference to an annual household wealth tax that the French government had instituted in 1945.

Austerity produced smaller, lighter cars with less powerful engines, but there were other reasons for the difference. Car ownership was less widespread than in the United States, and many people considered themselves fortunate to own a bicycle, or maybe a scooter or motorcycle—Britain alone manufactured four million bicycles annually. The tight streets of old European cities required narrow cars that only seated two across. Driving conditions were different, too. Writing in 1953, Maurice Olley, a British automotive engineer who worked for Rolls-Royce and later on Chevrolet's Corvette, described the differences: "The British motorist doodles along his lovely country lanes, admiring the scenery and dodging almost subconsciously the bicycles, children, dogs, wheelbarrows, flocks of sheep and all the thousand assorted hazards of British motoring, with as much pleasure as the Continental driver in his noisy minicar tears around the curves of his bumpy *route national*, or as the American sets out on his superb tollroads in a vast boudoir on wheels, strolling along at an effortless 70 mph."

Britain, France, Germany, and Italy had the highest rates of car ownership, and each country had a large number of low-volume manufacturers—more than two dozen in England. By comparison, by the end of the 1950s there were effectively only five American car companies: the Big Three plus Studebaker and AMC. The largest European car makers such as Fiat and Renault were small by American standards, although this did have advantages. Innovation, which was driven by intense competition, was less costly, and small enterprises, often family-owned, tended to be more individualistic and often more adventurous than large corporations. Austerity precluded the costly American practice of introducing annual model changes, and cars developed in the 1930s, such as the French Traction Avant and the Italian Topolino, remained in production virtually unchanged well into the 1950s.

Although some European car manufacturers exported their products, they faced trade barriers and their chief markets were local. National tastes varied. The French liked four doors, even in very small cars; German buyers favored aerodynamic bodies to suit the high-speed autobahns; and the British tended to prefer traditional forms. The overarching influence of American styling also varied. According to Olley, American trends had the least effect on German car design, were halfheartedly resisted by the French and the British, and were embraced by the Italians, although in a scaled-down form.

It is generally agreed that the best small European cars of the postwar period were French. The leading companies were Citroën, Renault, Panhard, and Peugeot. We have already seen how Citroën developed a budget car geared toward rural users. In 1946, two years before the Deux Chevaux appeared, Renault launched a small car that it had developed secretly during the war. The company's design director was Fernand Picard (1906–1993), an engineering graduate of an elite *école normale supérieure.* Like the Volkswagen, the Renault 4CV had a rear engine and rear-wheel drive, although the engine was water-cooled and the slightly shorter body had four doors—with sliding windows. The Renault's monocoque body was lighter than the Volkswagen, and hence got by with a smaller power plant, a 760 cc four-cylinder engine that produced a top speed of fifty-eight miles per hour. The 4CV had lightweight seats like the Citroën 2CV, but it was in many ways a less original body design, a miniature version of an American coupe of the 1930s. The 4CV was popular with buyers, however, and remained in production until 1961. The stalwart engine had an even longer life, since it was successfully reused in the Renault 4L, the little car that Ralph and I drove from Paris to Athens.

"The root trouble with American automobiles," observed a 1957 issue of *Motor Trend,* "would seem to be that they are now designed primarily by stylists and fashion experts, while European cars are built by engineers and drivers." The latter practice was established early. René Panhard (1841–1908) and Émile Levassor (1843–1897), who had founded the first French automo-

Renault 4CV, 1947

bile company in 1887, were both graduates of the École Centrale Paris, France's leading engineering school. Their company was responsible for many innovations, such as the Système Panhard, and the Panhard rod, which prevented lateral movement of the axle and is still in use. Panhard et Levassor cars became dominant in motor racing—Levassor himself died in an accident while driving in the Paris–Marseilles–Paris rally.

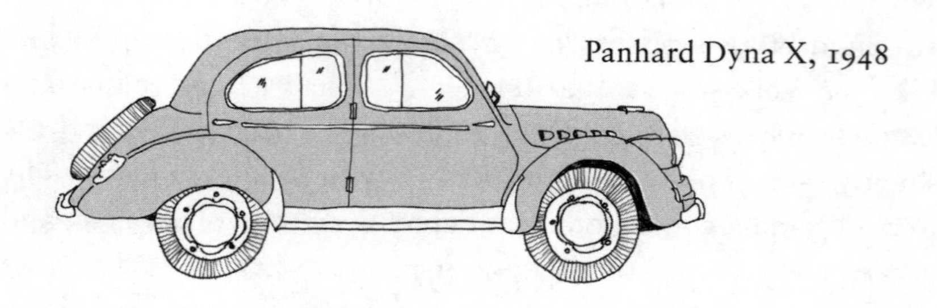

Panhard Dyna X, 1948

During the 1930s, the renamed Panhard company was known for its large, technically advanced luxury cars, but after the Second World War expensive cars were no longer in demand and the company turned to more economical models. The design for its first small car was licensed from Jean-Albert Grégoire (1899–1992), a freelance engineer who was a devotee of front-wheel drive. The resulting car, the 1948 Panhard Dyna X, was a four-door sedan similar in size and overall appearance to the Renault 4CV, but with a crucial difference: the engine was in the front and drove the front wheels. The body was a lightweight aluminum alloy—another Grégoire preoccupation. Like other small French cars, the power plant of the Dyna was not much larger than a motorcycle engine. Thanks to Louis Delagarde (1898–1990), the talented Panhard engineer in charge of engine design, the 610 cc, air-cooled two-cylinder boxer powered the lightweight car to a top speed of sixty-eight miles per hour. The Dyna X was available in two- and four-door sedans, a station wagon, and a van. The Dyna had a spacious interior, low fuel consumption, and effective performance, but it was not a very attractive design, and thanks to its ornate trim, the car acquired the nickname "Louis XV." The Dyna was outsold by the plainer and less expensive Renault 4CV, which benefited from a larger dealer network and became the first French car to sell more than a million.

The first British car to pass the one million mark was built by Morris Motors. The founder, William Morris (1877–1963), was a self-taught mechanic who had started building bicycles and motorcycles, and by 1912 was building automobiles. Thanks to the use of American-style assembly lines, Morris Motors became the largest car company in Britain, and manufactured a range of cars that included a small four-seater that competed successfully with the Austin Seven. During the 1930s, William Morris (now Lord Nuffield) acquired several smaller companies, including Wolseley, known for its luxury cars; Riley, one of the oldest British car companies; and MG, a maker of two-seater roadsters. In 1941, despite wartime conditions, Morris decided to start development of three new models that would be launched once the war was over; the models were small, medium-size, and large cars. The man put in charge was a young talented engineer, Alec Issigonis. Alexander Arnold Issigonis (1906–1988) was born in the port city of Smyrna. His Greek father, a naval engineer, had studied in London and become a British citizen, and during the Greco-Turkish War the family was evacuated to Malta and eventually settled in England. The young Issigonis, who studied mechanical engineering, worked for Austin and later joined Morris.

The small car that Issigonis designed, which would become the company's longest-selling model, was the Morris Minor. It had a front-mounted engine that drove the rear wheels, which sounds conventional, but like Ferdinand Porsche, Issigonis believed that

Morris Minor, 1948

a budget car should incorporate advanced engineering. Issigonis was a racing enthusiast—in the 1930s, he had successfully built and raced his own car—and he wanted the new Morris to provide good roadholding and safe handling, especially as many of the car's owners would be first-time buyers with little driving experience. At a little over twelve feet, the Minor was small, but Issigonis, who had studied German and Italian budget cars, wanted something better than another "shoehorn car." "People who drive small cars are the same size as those who drive large cars, and they should not be expected to put up with claustrophobic interiors," he observed. To maximize interior space in the cabin, he located the engine as far forward as possible—actually, ahead of the axle. He also used fourteen-inch wheels instead of the usual eighteen-inch, which reduced the size of the rear-wheel wells, while also improving handling. The wheels were pushed out to the corners of the body for the same reason. Despite the car's short length, the seats were entirely between the axles, producing a more comfortable ride. The car was fifty-six inches wide, which compared favorably to the Renault 4CV and the Panhard Dyna X, but at the last minute Issigonis widened the body even more, adding four inches. This enlarged the interior and provided more stability and better handling, giving the Minor an impressive stance.

Issigonis did not have a supportive boss. Lord Nuffield looked askance at "that foreign chap" and his radical ideas, and he would have preferred modifying a prewar model over a new design. He vetoed several of Issigonis's innovations, such as a rear torsion bar suspension instead of conventional leaf springs, and an ingenious 800 cc boxer engine that could have been easily upgraded to 1,100 cc for the export market (about a third of the first year's production would go for export, mainly to Canada and Australia). Instead, the Minor made do with an existing prewar engine that produced a top speed of sixty-four miles per hour.

The Morris Minor was unveiled in 1948. What had started as a small budget car had become a roomy family car. A family car with *character.* Issigonis believed that a small car should be aesthetically attractive and not simply cheap. A skilled draftsman, he was responsible for styling the Minor, which, unlike most British cars of the prewar period, was not boxy. Like the small Renault and Panhard, the body design included accentuated fenders in the front and an aerodynamic rear. The streamlining was emphasized by the low, flush-mounted headlights and the smooth grille. There was just enough chrome trim to lend the car an air of distinction.

The widened Morris Minor was much smaller than the smallest American car, but in postwar Europe it represented a step up from an entry-level budget car, and today it would be called a compact car. The same year that the Minor was introduced, Ford's German subsidiary launched the Taunus G73A, a two-door four-passenger sedan, slightly longer and heavier than the Minor. The Taunus was a revived prewar model that had been manufactured by Ford in Cologne from 1939 to 1942, when its production was halted by the war. The streamlined Taunus looked American because it *was* American—it had been designed in Detroit by Bob Gregorie, who was responsible for the Lincoln-Zephyr. At the time, Gregorie was also working on the 1939 Ford Tudor, an inexpensive entry-level family car, and the two designs had many features in common: large front fenders, flush headlights, and a streamlined, sloping rear. The Taunus had a similar silhouette to the Tudor, but on a considerably shorter chassis, and the shrunken rear gave rise to the nickname *Buckeltaunus*—"humpback Taunus." The German car got a smaller engine—a 34-horsepower four-cylinder that produced a top speed of 65 miles per hour, compared to the Tudor's 60-horsepower V8 that produced a top speed of 120 miles per hour. In other respects, the Taunus and the Tudor were similar except that the former lacked running boards and had less chrome trim and a simpler grille—just slots.

Ford Taunus G73A, 1948

The American influence was felt in another midsize car. The venerable French company Peugeot had started in 1810 as a family-owned steel foundry manufacturing tools, bicycles, and kitchen appliances (including pepper mills that are still sold today). Armand Peugeot (1849–1915)—like Panhard and Levassor, an École Centrale engineering graduate—expanded the business into automobiles, building his first internal combustion vehicle in 1890. An early popular car was the Bébé, a small two-seater only nine feet long. In time, Peugeot became one of the leading French car companies, also known for its racing victories. In the interwar period, Peugeot made large, powerful cars, midsize cars with smaller engines, and small cars such as the 1929 Peugeot 201, which was then the least expensive car on the French market. These cars tended to be aerodynamically designed and are often compared to the Chrysler Airflow.

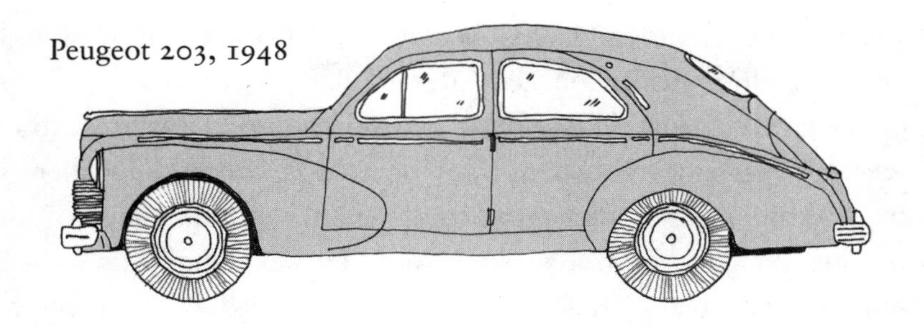
Peugeot 203, 1948

The Peugeot 203 arrived in 1948. It had an aerodynamic body similar to the Taunus, although it was slightly longer and heavier. The roomy four-seater was powered by a 1.3-liter four-cylinder engine that produced a top speed of seventy miles per hour. The rugged and reliable Peugeot proved itself a successful rally car, with good handling thanks to its independent front suspension and rack-and-pinion steering. With its streamlined styling, bulging fenders, and sloping rear, the 203 resembled a smaller version of the Stylemaster, an entry-level Chevrolet introduced two years earlier. Chrome trim in both cars was limited to the grille, bumpers, and a long waistline stripe. Although the five-passenger Chevrolet weighed almost a thousand pounds more than the Peugeot (which had a monocoque body), and the American six-cylinder engine put out ninety horsepower, the Stylemaster was only slightly more expensive than the smaller French car, attesting to the advantages of American mass production. On the other hand, due to "dynamic obsolescence," the Chevrolet Stylemaster had a production life of only three years, whereas the Peugeot 203 remained in production until 1963—fifteen years—and in the long run actually outsold the Chevy.

As the French postwar economy improved, Panhard, seeing sales of its small Dyna X declining, needed a midsize model that could appeal to the burgeoning family-car market and compete with the Peugeot 203. The new model had a front engine and front-wheel drive, and because the company was not in a financial position to manufacture an entirely new car, the engine, transmission, steering, and suspension all came from the Dyna X. There the similarity ended. Louis Bionier (1898–1973), a self-taught engineer who was in charge of design at Panhard, had been conducting wind tunnel experiments, and the 1954 Dyna Z (there was no Y) was a smoothly sculpted car with a drag coefficient of 0.34. Bionier designed a monocoque body that was made of aluminum on a tubular-steel subframe. The extremely light body made it possible to reuse the small two-cylinder boxer engine, whose displacement Delagarde had increased to 851 cc, producing a top speed of eighty miles per hour. This was particularly impressive because the Dyna Z was a large car: although it had the same wheelbase as the Peugeot 203, it was longer and wider and could accommodate five passengers.

Panhard Dyna Z, 1954

While the body design of the Peugeot 203 was a throwback to prewar American sedans, Bionier's striking Dyna Z did not take its styling cues from any previous car—it was an original. There were no fenders; the windshield and the rear window were set in rubber gaskets instead of the usual metal frames; the turn signals were located high up near the doors; there was little trim, except for a single chrome stripe at the foot of the doors. On the interior, the dashboard, which was padded, was bereft of instruments; instead, the speedometer, fuel gauge, and a cluster of knobs and rocker switches were in a small pod behind the steering wheel. The most striking feature was the front of the car. Since the engine was air-cooled, there was no radiator, and instead of a conventional grille, Bonier designed a mouthlike opening with a central fog lamp. The appearance of a cheerful face was emphasized by the flush headlights. It was altogether an unusual car.*

COMPACT EUROPEAN CARS such as the Morris Minor, the Ford Taunus, and the Peugeot 203 were essentially small versions of American cars. There was one category of postwar European automobile that had no American equivalent: the microcar. An early example was the British Bond Minicar. Lawrence Bond (1907–1974) was a self-taught mechanic whose small workshop in Lancashire had produced aircraft parts during the war. He built small racing cars powered by 500 cc motorcycle engines, a popular British racing class. In 1948, he devised a small three-wheeled car for his wife, something she could use for shopping errands. Bond did not have the resources to commercialize his idea, and he sold the manufacturing rights to a small local machine shop that refined the design and the following year launched the Bond Minicar. Why three wheels? In the immediate postwar era, the British sales tax on automobiles approached fifty percent, but a three-wheeler was classified as a motorcycle and was taxed at half that rate. Registration and insurance were less expensive, too, and the operator of a three-wheeled vehicle did not require a driver's license.

The Bond Minicar was a two-passenger vehicle with a small engine driving the steerable front wheel, and with two wheels in the rear. The motorcycle engine was a two-stroke—that is, it was lubricated by simply adding oil to the fuel. Over the next seventeen years, there were seven consecutive Minicar versions, with enlarged

* The innovative Dyna Z was not an unqualified success. Because of faulty cost estimates and a rise in the price of aluminum, Panhard was obliged to switch to sheet steel for most of the body, which added about three hundred pounds to the weight of the car and required modifications to the suspension. The financially strapped company was obliged to take Citroën on as a minority partner. Over time, Citroën increased its stake in the company until it had a controlling interest, and in 1965, it terminated the Panhard marque.

engines, improved suspensions, rack-and-pinion steering, and restyled bodies. The Mark B—the second iteration, which was produced in 1951—can serve as an example. The 197 cc single-cylinder engine produced eight horsepower, about the same as a large modern walk-behind mower, and just as noisy. The manual transmission had three speeds, although no reverse gear, this being originally a motorcycle engine. The little engine, cantilevered from the front wheel assembly and linked by a chain drive, produced a top speed of about fifty miles per hour. This speed was possible because the aluminum body was extremely light—the whole car weighed only 420 pounds. The front wheel assembly was mounted on a hydraulic shock absorber; the rigid rear axle rode on a sliding block controlled by coiled springs. The car was nine feet long, and the large engine compartment was necessary to provide space for the swinging engine assembly. The padded bench seat could accommodate three in a pinch. There was a storage space behind the seat; some later models were lengthened to accommodate rear seating for children. The Mark B had a fold-down canvas top with detachable side screens. There were no doors—the car was so low, you just stepped in.

Bond Minicar
Mark B, 1951

Over time, there were many efforts to give the Minicar the appearance of a conventional automobile by the addition of side doors, mock fenders, chrome trim, an openable trunk, even abbreviated tail fins. But the fact remained that there is something unsettling about a car whose front is precariously balanced on a single front wheel. Unlike the Citroën 2CV, which exploited its unusualness, the idiosyncratic Bond Minicar remained an unresolved design: a three-wheeled vehicle that really wanted to be a four-wheeler.*

This was not the case with the Italian Isetta, a distinctly odd car that embraced its oddity. The manufacturer was Iso, an Italian company that had been founded in 1939 by Renzo Rivolta (1908–1966) to make refrigerators. After the war, Rivolta, an engineer, branched out into motor scooters and motorcycles. He also produced a three-wheeled tricycle truck, basically a scooter front with a two-wheeled utility bed behind. In 1952, Rivolta had the idea of developing a passenger version. He turned to a professor at the nearby Milan Polytechnic to design the car. Ermenegildo Preti (1918–1986) was an aeronautical engineer who had experience designing civilian and military gliders. He had been toying with designs for small cars, which may be why Rivolta approached him, but Preti had no automotive background, which explains his unorthodox solution. The small wooden model that he brought to his first meeting with Rivolta had three wheels—two in the front and one in the rear—and an unusual body. It has been said that Preti based his design on a glider cockpit. That story may be apocryphal, but it well describes the glazed egg-like shape that gave rise to the nickname for later micro vehicles: bubble cars.

How does one get into a bubble? Unlike virtually every automobile that had ever been built, the Isetta was not entered from the side, but from the front—a solution that Preti had used in a military transport glider. The entire front of the car, including the windshield, was a side-hinged hatch. Preti and his young colleague Pierluigi Raggi (1924–2012), also an aeronautical engineer, were used to designing tight cockpits. When the hatch was opened, the attached steering wheel and instrument pod swung out of the way. The wide bench seat could accommodate

* About fifty-five thousand Bond Minicars were sold between 1949 and 1966.

two adults and a small child; a parcel shelf behind the seat covered the engine. As in an aircraft cockpit, full glazing provided 360-degree visibility. The canvas roof folded open, to provide ventilation and also to serve as a second means of egress in case of a frontal collision.

Iso Isetta, 1953

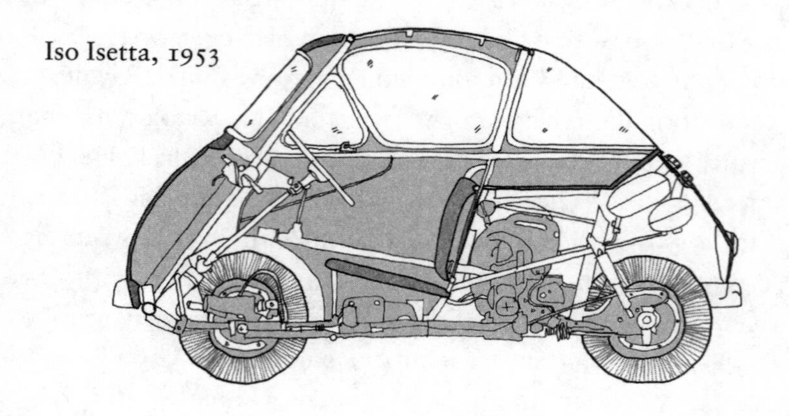

The Isetta was smaller than a Bond Minicar, although the profusion of glass made the Italian car heavier, which necessitated a larger engine. The 236 cc air-cooled single-cylinder two-stroke motorcycle engine produced nine and a half horsepower and a top speed of forty-seven miles per hour. The prototype three-wheeler proved unstable, and the single rear wheel was replaced by two wheels eighteen inches apart, close enough to dispense with a differential. The engine powered the small rear wheels (ten inches in diameter) by means of a chain drive.

Preti and Raggi did not design a miniature version of a conventional car. Instead, they started from scratch to rethink what a two-passenger microcar might be. The result was an automotive original with a distinctive sense of style. Nevertheless, the Isetta was not a marketing success; in two years, fewer than five thousand were sold. The reasons may have been Iso's lack of a dealer network, a finicky engine that required careful maintenance, and competition from Fiat's low-cost Topolino, which was still in production. Rivolta, a mercurial businessman who went on to manufacture luxury sports cars, ceased production of the Isetta after two years, but continued to license the car internationally, in Britain, France, Germany, Spain, Brazil, and Argentina.

The most successful licensee was BMW in Germany, which kept the original body design but retooled and upgraded most of the mechanicals, especially the problematic engine. The combination of German know-how and Italian flair proved a success, and over a production life of seven years, BMW would sell more than 160,000 Isettas. Microcars were particularly popular in postwar Germany; the aircraft companies Messerschmitt and Heinkel both produced models, though neither provided serious competition to BMW's Isetta. In Britain, scores of small companies built a variety of unconventional three-wheelers. Microcars were given a boost in 1956 by the Suez Crisis, which interrupted oil supply and resulted in shortages, higher prices, and rationing. Overnight, fuel consumption became a major concern of car buyers. The Bond Minicar, still in production, could do seventy miles on a gallon of gas; the BMW Isetta, which was available in Britain, could do better than ninety.

"God damn these bloody awful bubble cars," fumed Leonard Lord (1896–1967), head of the British Motor Corporation (BMC), which had been founded in 1952 through a merger of Morris and Austin. BMC was the third-largest car company in the world, with a wide range of cars but no models small enough to compete with the microcars. "We must drive them off the road by designing a proper miniature car," Lord told his deputy engineering director, Alec Issigonis—the pair had earlier worked together at Morris Motors. Lord gave Issigonis specific instructions: the four-passenger car should be four feet wide and ten feet long, which was about the size of the new Fiat Cinquecento, launched that year; the passenger compartment should occupy sixty percent of the length; and the engine should be an existing BMC product.

Issigonis's solution was simplicity itself: two boxes, one for the cabin and a smaller one for the engine, anticipating the design of the Renault 4. He gave the car front-wheel drive, like the Morris Minor, but turned the engine ninety degrees (as Giacosa had done with the Fiat 500) so that it was parallel to the axle, an arrangement that was both mechanically simpler and more compact. The modified Morris Minor engine—848 cc, four cylinders, and water-cooled—produced thirty-three horsepower and a top speed of seventy-five miles per hour. The radiator was on the side, and the four-speed transmission, which was lubricated by engine oil, was located in the sump, which further reduced the size of the engine compartment. Another innovation was the suspension: compact

BMC Mini, 1959

rubber cones instead of springs. The tiny wheels (ten inches in diameter), for which Dunlop developed special tires, increased interior space, improved handling, and facilitated steering (which was not power-assisted). As a result of these engineering changes, a remarkable eight feet of the car's length was available for the cabin, bettering Lord's specification.

The monocoque steel body—designed by Issigonis—was a case of form following function. He refrained from any styling flourishes—no aerodynamic curves, no trim, virtually no chrome at all. Economy guided many of his decisions: external hinges on the two doors and the trunk, sliding front windows, and hinged rear windows that could only be cracked open. The interior was spartan. The only instrument was a large speedometer with a built-in odometer and fuel gauge; the tiny pod included two knobs (heater and choke), two switches (lights and wipers), and the ignition. A wide shelf replaced the usual dash, additional storage being provided in door pockets and in optional wicker baskets that fitted beneath the seats. A parcel shelf, which served to brace the body, covered the trunk, which was accessible from outside. The bottom-hinged trunk lid could be left open to provide extra space—shades of the old woodie wagon.

The selling price of the Mini, which initially was marketed as Morris Mini-Minor and Austin Seven, was just under £500 (about $13,000 today), and the car was inexpensive to operate, with fuel consumption of fifty miles per gallon.* With its small wheels pushed out to the corners and a low center of gravity, the handling and roadholding were excellent and the nimble car was a pleasure to drive. Thanks to these qualities, like the Volkswagen Beetle, the Citroën 2CV, and the Fiat Cinquecento, the Mini overcame the highest hurdle for a budget car: being thought of simply as "a cheap car." The Mini had that indefinable something that distinguishes a classic. The ageless styling proved an advantage, because the car remained in production for forty years—a total run of 5.3 million cars.

The drawback for the manufacturer of an inexpensive car like the Mini is the small profit margin, and in 1962 BMC introduced a higher-priced family-size version, also designed by Issigonis. The

* The inexpensive Mini killed off the bubble car, as Lord intended. BMW discontinued the Isetta in 1962, the Messerschmitt was retired in 1964, and the Bond Minicar in 1966.

1100, which had a similar transversely mounted engine powering the front wheels, did not become legendary like the Mini, but it is an interesting design. "We have tried to produce a good looking, functional car—while cutting out as far as possible the risk of things going wrong. My main plan was to design a motorcar to travel as efficiently as possible from A to B, with full comfort over really rough roads," said Issigonis. The "bigger and better Mini" was not simply roomier, but included such innovative technical features as disc brakes. The 1100's unusual "Hydrolastic" suspension system was designed by Alex Moulton (1920–2012), a Cambridge-educated engineer who had been responsible for the Mini's rubber cone suspension.* Moulton was inspired by the pliant suspension of the Citroën 2CV, but instead of torsion bars, he devised a system of interconnected rubber balloons filled with a mixture of water and alcohol that enabled the car to maintain stability over bumps without the excessive roll of the Citroën. The long wheelbase allowed the 1100 chassis to serve as a four-door sedan and a station wagon, as well as a two-door coupe. Like the Mini, the small wheels—twelve inches in diameter rather than ten—were pushed to the corners. Although the wheelbase was the same as a VW Beetle, the 1100 was a foot shorter, reflecting Issigonis's view that what the customer wanted was a large cabin but a small car.

The 1100 had a sloping fastback, with a lower hatch giving access to the trunk. The elegant body was the work of Carrozzeria Pinin Farina of Turin, the company founded by Battista "Pinin" Farina (1893–1966), who had previously done work for BMC. Farina was responsible for such classics as the Siata Daina and the Alfa Romeo Giulietta Spider, and the combination of British engineering and Italian design was particularly successful. Versions of the 1100 appeared under several BMC marques, including Morris, Austin, and MG, the latter being exported to North America. In 1965, I drove with a friend in her MG 1100

* In 1965, Moulton designed a revolutionary small-wheeled bicycle with high-pressure tires and a front and rear suspension. I bought one that summer.

Morris 1100, 1962

from Montreal to Percé on the Gaspé Peninsula, a distance of more than six hundred miles, and I remember the interior as surprisingly spacious. The 1100 remained in production from 1962 until 1974, and for much of that time it was the best-selling car in Britain; more than 2.2 million were made, not counting the licensed offshore production in Belgium, Spain, Italy, Yugoslavia, South Africa, and Australia.

If the Mini had a longer run than the 1100, that may have been because it became a popular rally and club racing car. In 1961, John Cooper (1923–2000), who had been a successful Formula One race car driver and was a friend of Issigonis, convinced BMC to produce a performance version of the Mini. With Hydrolastic suspension, disc brakes, and a larger engine with twin carburetors, the Mini Cooper had a top speed of ninety miles per hour. I owned a used Mini Cooper in the late 1960s. Driving it was a different experience than my previous car—the Volkswagen that had taken me from Hamburg to Valencia. It was not just a question of speed, but of what drivers call "road feel." And sound: at higher speeds, the carbureted engine produced a satisfying roar. Maroon with a white roof, my car had obviously been raced—the hand brake had been replaced by a fire extinguisher. Of course, a retired racing car was a rash purchase. I spent a year fussing with mechanical problems—with a friend's help, I bled the brakes and replaced the pads—and I finally traded it in for the red Citroën Deux Chevaux.

"It'll still be fashionable when I'm dead and gone," said Issigonis of the Mini. He was still alive, although retired, when I had my second encounter with a Mini, in 1984. I used to get my car, at that time a Subaru station wagon, serviced at a garage run by an Italian acquaintance of my father. During one visit, I noticed a new car on display that looked almost exactly like my old Mini. What's that? I asked. "An Innocenti," was the answer. "We've started selling them." This little car was Italian! It was a convoluted story. Innocenti was a Milanese company that, during the postwar period, manufactured Lambretta motor scooters and eventually made cars under license from BMC, including the Mini and the 1100. The cars sold well, and in 1974, British Leyland, which had taken over BMC, bought the Italian company. As part of an ambitious planned expansion, British Leyland hired Gruppo Bertone, a leading *carrozzeria* led by Nuccio Bertone (1914–1997), to freshen up the Mini for the Italian market. The designer in charge, Marcello Gandini (1938–2024), had been responsible for high-performance Lamborghini sports cars such as the beautiful Miura and the Countach, but his work on the Mini was restrained. He kept the overall dimensions: same wheelbase, same height, a few inches longer and wider, and a little more angular body. The biggest difference was a larger top-hinged rear hatch that provided more convenient access, and a folding rear seat that created a large rear cargo space. The interior was nicer than in the spartan British original, with wind-down windows and quarter-glass vents. The pretty body had the same no-frills aesthetic, except for one odd detail: Gandini gave the trailing edge of the roof a tiny upward flip, the barest hint of a spoiler, as if to say, "I could have been fancier."

Innocenti
Mini 90L, 1974

A year after the restyled Innocenti Mini was launched, British

Leyland entered bankruptcy and the Innocenti company was bought by Alejandro de Tomaso (1928–2003), an Argentinian entrepreneur. An ex–racing car driver, de Tomaso had founded an Italian car company based in Modena that owned Maserati and built limited-production sports cars such as the Mangusta and the Pantera. Perhaps seeking to diversify, de Tomaso continued to manufacture the Gandini-styled Mini, although he improved the suspension and substituted a more reliable Japanese engine—a peppy three-cylinder from the best-selling Daihatsu Charade. The Innocenti Minitre sold well in Italy, and in the 1980s it was exported to France, Belgium, Switzerland, and Germany, arriving in Canada in 1984. That's when I saw it—and bought it.*

My wife, Shirley, patiently tolerated my all-too-frequent car purchases. She was born and raised in downtown Montreal and had never owned a car—taxis were more her style. We met in the late 1960s, when I had the red Deux Chevaux and she scooted around town on a VéloSoleX. The motorized bicycle, a French invention of the 1940s, had a tiny (49 cc) gas motor with a roller resting on—and powering—the front wheel; with the roller disengaged, the moped became a conventional bicycle.† Shirley's Solex was the first in the city—she had convinced the dealer to give her one; he figured that an attractive young redhead driving the new import would be good publicity.

Shirley had a driver's license, but she didn't particularly enjoy driving, so I was surprised at her reaction to the Innocenti—she loved it. I think it was the combination of small size, Mini handling, and Italian design that appealed to her. The driving experience had less to do with speed—although the car could reach about ninety miles per hour—but rather with the way it made you feel. In a word, it was fun.

* In 1990, Fiat acquired Innocenti and ceased production of the Minitre three years later.

† In the opening scene of the 1975 film *Three Days of the Condor*, the character played by Robert Redford is shown riding a Solex to work in Manhattan.

CHAPTER
EIGHT

FUN

If the value of a car consists of practical values and emotional appeal, the sports car has very little of the first and consequently has to have an exaggerated amount of the second.

—ZORA ARKUS-DUNTOV

MY FOURTH CAR, AFTER THE VOLKSWAGEN, THE MINI Cooper, and the Citroën, was another German. Today, BMW is a successful manufacturer of upscale cars, but in 1962, when the model I bought was introduced, the Bavarian company was emerging out of near bankruptcy and staking its future on a brand-new vehicle. In the 1930s, Bayerische Motoren Werke had been known for building large, expensive cars that competed with Daimler-Benz. During the war, BMW manufactured aircraft engines and motorcycles, and it was not until 1952 that the company resumed car production. Its first postwar car was the 501, a luxury sedan that acquired the nickname *Barockengel*—Baroque angel—for its curvaceous styling. There was little demand for such a car in the recovering West Germany, and the 501 sold poorly. Shortly after, BMW acquired the Isetta franchise, but an inexpensive bubble car was not an adequate lifeline, and by 1959 the struggling automaker was about to be taken over by its old rival Daimler-Benz. Only a last-minute investment by a leading German industrialist and minority shareholder saved the company.

In 1962, BMW launched what it called Neue Klasse (New Class), a line of smallish sedans and coupes that emphasized build quality and performance. The first model was a four-door, four-passenger sedan. The BMW 1500, which had a 1.5-liter engine, was a compact version of what the British call an executive car, targeted at professionals and middle managers, a category defined as "comfortable, refined and displaying some form of driving pleasure." The 1500's technology was up-to-date, though hardly revolutionary: a monocoque body, independent rear suspension, and front disc brakes. The front suspension used MacPherson struts, which consisted of telescopic dampers that provided both a steering pivot and an independent suspension for the front wheels and were simpler than the traditional double-wishbone suspension that had been used since the 1930s.* The conservative BMW body consisted of three boxes: one for the engine, one for the passengers, and one for the trunk. The boxes were upright rather than streamlined, producing a tall, roomy cabin and a capacious trunk.

Four years later, BMW introduced a two-door version that was slightly smaller (a foot shorter, five inches narrower, three hundred pounds lighter), with simpler interior finishes but a more powerful 1.6-liter engine producing ninety-four horsepower and a top speed of 100 miles per hour. This entry-level car, the BMW 1600, was exported to North America and turned out to be a milestone for the company. What made the 1600—and its successor, the 2002 (which had a two-liter engine)—special was that they combined performance and handling with practicality and comfort at a reasonable price.† The February 1967 issue of *Car and Driver* didn't mince words; it called the BMW 1600 "the best small sedan we ever drove."

The three-year-old BMW 1600 that I bought in 1972 was certainly the best sedan I had ever driven. The car was character-

* Strut suspension was invented in the late 1940s by Earle S. MacPherson (1891–1960), a Ford engineer, and was first used in the 1950s by British Ford.

† A new BMW 1600 sold for $2,500 in 1966.

ized by a sort of genial equanimity; it was eager to please and do whatever was asked of it, whether cruising the highway or scooting along winding country roads. I agreed with *Car and Driver* editor David. E. Davis Jr.'s characterization of the 1600's sibling, the BMW 2002: "Like a good sheep dog, it is ill-suited for show competition, only becoming beautiful when it's doing its job." I liked everything about my car, especially the understated interior. The gauges and control knobs were grouped behind the steering wheel with the speedometer and a clock—I later added a tachometer—in a small instrument cluster that left a long, useful shelf instead of a dashboard. There were no fancy chrome trim or wood finishes. The doors had roomy storage pockets and convenient armrests. The seats, covered in a weave-pattern vinyl, held you firmly in place, and the tall windows provided clear visibility all around.

The exterior was similarly low-key. My car was sandy beige, a color that BMW called Sahara. The front end had a small version of the split kidney grille that had been a company trademark since the 1930s; the rear had two circular taillights that echoed the headlights. The B-pillars behind the driver were extremely thin, making for a very airy cabin. The most distinctive design feature was the high beltline wrapped around the entire car, which recalled the American Corvair. At this time, BMW did not have a styling department—the person in charge of body engineering was Wilhelm Hofmeister (1912–1978), a mechanical engineer. I'm tempted to call the 1600 a well-designed tool, except that would be misleading. The small car had a je ne sais quoi that transcended mere functionality.

Some of the design features of the Neue Klasse series are credited to the Italian automotive designer Giovanni Michelotti (1921–1980), whom Hofmeister brought in as a consultant. Michelotti had his own studio and coachbuilding firm in Turin and had worked for many of the leading *carrozzerie* such as Bertone, Ghia, and Pinin Farina on striking Italian sports cars and grand tourers. Michelotti's influence is visible in the many subtle details of the Neue Klasse, such as the reverse slope of the grille—the

BMW 1600, 1966

opposite of most cars—and the deceptively simple undecorated body. When *Car and Driver* described the BMW 1600 as "an Alfa Romeo built by Germans," that was not hyperbole.

THE BMW 1600 WAS NOT the first sporty European sedan to be exported to North America; the Volvo PV444 arrived about a dozen years earlier. This rugged, winter-proof Swedish car was especially popular in Montreal, and in the seventies my friend Jean-Louis owned the 444's successor, the 544—the same body, but a slightly larger engine. His two-door sedan, which was about the same size as my BMW 1600, was ten years old. Actually, it looked much older than that if you went by the bulbous hood, prominent fenders, and humpback rear. The Volvo resembled a scaled-down version of the sort of cars that Humphrey Bogart drove in 1940s movies such as *The Maltese Falcon* and *High Sierra*.

The Volvo PV444 that arrived in North America in 1955 had a split windshield and rear window, but its performance belied its staid looks; it was fast. The Volvo engineers had teased seventy horsepower from the 1.4-liter four-cylinder carbureted engine, giving a top speed of ninety miles per hour. "No other sedan of under 1.5 liters we have ever tested has turned in acceleration times up to 60 mph equal to this Scandinavian import," observed the 1957 *Road & Track* review. The magazine described a recent road race in Pomona, California, in which a sleek Porsche 356 had finished first, and "there, hounding along in 2nd place overall, came a Volvo, comfortably ahead of all kinds of sports machinery." By then, the 444 had proven itself in European rallies and road races, and Volvo advertisements, which highlighted safety and reliability, were calling it a "family sports car."

The design of Jean-Louis's old-looking car really was old—it dated back to 1944. Volvo was a relative latecomer to car manufacturing. In 1926, businessman Assar Gabrielsson (1891–1962) and engineer Gustaf Larson (1887–1968) founded the company as a subsidiary of SKF, the giant Swedish manufacturer of ball bearings (Volvo became independent in 1935). Although Volvo started manufacturing cars in 1927, in its early years the company was best known for trucks and buses; passenger cars were produced in limited numbers for the local market.

Gabrielsson and Larson noticed that inexpensive German cars were popular in Sweden, and in 1947 Volvo launched its own mass-market car, a midsize two-door sedan. The Volvo PV444 (PV simply stood for *personvagnar*, or passenger car) had a conventional front engine and rear-wheel drive mated to a monocoque body, an arrangement that the Volvo engineers modeled on the 1939 Hanomag. The two-door German sedan had a streamlined body with a curving grille that recalled the Chrysler Airflow of a few years earlier. Volvo copied the general configuration and monocoque construction, but not the body design. The engineer responsible for the bodywork of the 444 was Carl Edward Lindberg (1900–1969), who had worked in Detroit for Studebaker, Cadillac, and the Briggs Manufacturing Company, so it is not surprising that the Volvo 444 resembled a prewar American coupe. The retro styling actually suited Volvo's later offbeat image, and it is probably what attracted Jean-Louis, who was a furniture designer. The dependable 444—which cost less than $2,000 in 1955—and its successor 544 were popular with

Hanomag
1.3 liter, 1939

Volvo PV444 LS,
1947

buyers outside Sweden. By the time production ceased in 1965, almost half a million of these cars had been sold and Volvo had become a global brand.

The other Swedish car manufacturer of the postwar period was Saab (Svenska Aeroplan AB), a company founded in the 1930s to build aircraft for the Swedish air force. The company's first car, the Saab 92, was introduced soon after the Volvo 444, in 1949. The two-door sedan was smaller, closer in size to the Volkswagen, which was then resuming production in Wolfsburg. Designed by a team of aeronautical engineers, the Saab was unorthodox in almost every respect. Like the Volvo, it had a monocoque body that was designed to be extremely stiff, like an airplane fuselage, and was reinforced at crucial points with thicker side sills and doors. The front was designed to be a crumple zone, and it was said that a Saab could roll over several times and resume driving. There were mechanical innovations, too: the generator doubled as a water pump, and the engine that drove the front wheels was mounted transversely.

The two-stroke engine of the Saab was patterned on the German DKW, a compact sedan. Today, two-stroke engines are commonly associated with chainsaws and lawnmowers, but in the early days of motoring, they were frequently used in small cars, and not only microcars like the Bond Minicar and the Isetta. A two-stroke engine had fewer parts and was cheaper to manufacture than a conventional four-stroke engine; it was also lighter and required less maintenance. In addition, a two-stroke engine produced more torque at higher revolutions than a conventional internal combustion engine, making for improved acceleration.* The Saab engine was mounted transversely—a decade before Issigonis adopted that arrangement in the Mini. The 764 cc two-cylinder engine produced twenty-five horsepower and a top speed of sixty-five miles per hour. A car powered by such a small engine needed to be light, and thanks to its monocoque construction, the Saab, despite having a slightly longer wheelbase, was seventy pounds lighter than a Volkswagen. The Saab interior was surprisingly roomy. The seats—individual seats in the front and a bench in the rear—were of tubular-steel construction to reduce weight. To save money, the trunk in the early models had no opening lid and was accessed from the interior. Like the Volvo 444, which included a "bed set" option that could sleep two and was stored under the rear seat when not in use, a plywood kit converted the rear of the Saab into a double bed.

For assistance with the body design, the lead engineer, Gunnar Ljungström (1905–1999), turned to Sixten Sason (1912–1967), who worked in Saab's drafting department. Sason had studied painting and sculpture in Paris, and like Flaminio Bertoni at Citroën,

* Two-stroke engines also produce more pollution, and they were eventually discontinued in cars.

Saab 92, 1949

he brought an artistic sensibility to the project; moreover, he had been an air force pilot and understood aerodynamics, and, thanks to his streamlined design, the sleek little Saab had an extremely low drag coefficient of 0.35.

The Saab 92 has been described as a cross between a Bugatti 57SC, a Tatra, a Citroën DS, and a flying saucer. The car was a remarkably original feat of advanced engineering and design; after all, the Citroën DS would not appear for another six years, and unlike the Tatra, the Swedish car was inexpensive—in 1949, it cost the equivalent of a thousand dollars. Six years later, the successor Saab 93 appeared, with improved brakes, a one-piece windshield, an openable trunk, a wider interior that sat three in the rear, and a more conventional four-stroke three-cylinder engine that produced thirty-three horsepower. The 93 was Saab's first export model, and it proved extremely popular. *The Motor*, a British magazine, wrote: "It is a small family saloon of useful carrying capacity, plainly finished but fully equipped, docile to drive and economical to run. It is also a car which a motorist of sporting tastes can drive really hard without danger, and without any impression that the car is being punished." *Of sporting tastes*—like the Volvo, the Saab proved to be an exceptional road racing and rally car. The term "sports sedan" was not yet in common use, but these Saabs and Volvos were early examples, anticipating the BMW 1600: everyday cars that were also fun to drive.

EUROPEAN MOTORSPORT HAS A long history. The first road race, the Targa Florio, which made a three-lap, ninety-three-mile circuit of the island of Sicily, took place in 1906. The 1,167-mile Prinz Heinrich Tour, in which Ferdinand Porsche drove an Austro-Daimler to victory, was first held in 1908. In the early 1920s, a number of road races were inaugurated, such as the 24 Hours of Le Mans in France and the Mille Miglia in Italy. These endurance events, which took place on regular roads running through villages and towns, enthralled the public. Later, dedicated racetracks were built at Monza, north of Milan, and Nürburgring in the Rhineland. These were complicated layouts with twisty curves, hairpins, and hills. Early road races were restricted to production cars, which heightened public interest.* It was not that every European motorist was a weekend racer, but for many enthusiasts, the two-seater car, increasingly called a sports car, represented the paragon of motoring.

One of the classic cars of postwar motorsport appeared in 1948. It was the first automobile to carry the Porsche badge, although it was not the product of the famous father, but of his son. Before the war, Ferry Porsche (1909–1998) had run the family design consultancy in Stuttgart while his father oversaw production of KdF-Wagens in Fallersleben. In 1945, Ferry and his father were imprisoned by the French, who demanded a large cash payment for their release. Initially, the family was able to raise money only for the son, who, barred from returning to Stuttgart, had set up shop in the small Austrian town of Gmünd. His first commission was a racing car for Cisitalia, an Italian builder of racing cars and crafted sports cars. Porsche's design proved too complicated for the small company, and the car did not proceed beyond a working prototype, but the advance payment allowed

* Formula One, which described purpose-built single-seater racing cars rather than production cars, originated in 1946.

Ferry to arrange the release of his father and to develop a sports car on his own account.

Cisitalia had earlier built a small competitive racing car, the D46, designed by Dante Giacosa using parts from the Fiat Topolino. This may have inspired Ferry Porsche to try something similar using the Volkswagen. At this time, his own car was a VW convertible with a supercharged engine. "I saw that if you had enough power in a small car, it is nicer to drive than if you have a big car which is also overpowered. And it is more fun," he said. "On this basic idea, we started the first Porsche prototype." The Porsche 356 ("356" was the project number), a two-seater roadster with an air-cooled rear engine, shared many parts with the Volkswagen and had a similar construction, although it was a foot shorter and a foot lower. The goal was to make the car as light as possible, and the small boxer engine as powerful as possible. The body was initially handcrafted aluminum, though this was later changed to steel when production ramped up. The engine was basically a tuned 1.1-liter Volkswagen engine putting out forty horsepower; later, more powerful 1.3- and 1.5-liter engines were used. The minimalist body design was the work of Erwin Komenda, who had been responsible for the original KdF-Wagen. He produced a beautiful car with a smooth body, no extraneous trim, no grille, sunken oval headlights, and a streamlined back. Remarkably, this first Porsche included many of the trademark design features that would characterize the company's models for decades to come.

The 356 was unveiled at the 1949 Geneva International Motor Show, and although it was well received and resulted in orders from Switzerland and Sweden, during the first two years only fifty cars were built. Production increased after the company relocated to Zuffenhausen, near Stuttgart. Initially, sales were limited to Germany and Austria, but as word of the car spread, and especially after it began to be exported to the United States, sales continued to grow. This first models were a hardtop coupe and a convertible, hand-built by Reutter, a leading Stuttgart coachworks that had

Porsche 356A, 1955

built KdF-Wagen prototypes before the war. The Porsche 356 was not a cheap car—it cost three times as much as a Volkswagen—but it was well built and fast. And it was definitely fun.

WHEN I WAS COMING of age in the 1960s, I associated "Made in England" with quality: Raleigh bicycles, Wedgwood china, Burberry raincoats. My first car had been a Volkswagen, and I valued my Leica camera and Braun slide projector, but it was British sports cars that turned my head: the classic MG, the streamlined Triumph and Austin-Healey, and the eccentric Morgan. One of my professors owned the only Aston Martin in Montreal. But the quintessential British sports car was the Jaguar, and the quintessential Jaguar was the XKE. Jaguar had been founded in 1922 by William Walmsley (1892–1961) and William Lyons (1901–1985). They were neighbors on the same street in Blackpool—Walmsley was the tinkerer and Lyons the entrepreneur. Both were motorcycle enthusiasts, and they founded the Swallow Sidecar Company. Their handsome, streamlined aluminum sidecars sold well, and the business expanded into an all-purpose coachworks, eventually building car bodies using the chassis and engine of the Austin Seven, the "Baby Austin." The sidecar company grew into the SS Car Company. Walmsley stepped down, and the ambitious Lyons built larger and faster models: a 2.5-liter sports sedan and a stylish 3.5-liter open two-seater, both models named Jaguar. In 1945, to avoid the unfortunate association with the Nazi SS, the company was renamed Jaguar.

Prewar Jaguars used chassis and engines manufactured by others, but the company planned to unveil its own engine, a powerful 3.5-liter inline six-cylinder with dual overhead camshafts capable of 160 horsepower, at the 1948 London Motor Show. The so-called XK engine, designed by William Heynes (1903–1989), Jaguar's chief engineer, was intended for a forthcoming sedan that was aimed at the export market. Since the sedan was not ready in time for the London show, Jaguar hurriedly assembled a two-seater—not exactly a concept car, but rather a display car for the new engine, with the option of a limited production run in the future if there was interest. The car turned out to be so popular that Jaguar put it into full production as the XK120. Initially, it was available only as an open roadster; later, a convertible and a hardtop coupe were added. The "120" referred to its top speed, making the Jaguar roadster the fastest production car of its day.

The XK120 was the work of Heynes, William Lyons, and Fred Gardner (1912–1990), who was the longtime head of Jaguar's carpentry shop (the sidecars and early Jaguars had used wooden frames). Gardner translated Lyons's rough sketches into life-size wire models, and ultimately into wooden armatures on which aluminum body panels could be hand-formed. According to Lyons, the basic body shape was arrived at in a few weeks. Although the body was inspired by the 1936 BMW 328, an open two-seater, the Jaguar's smooth, sculptural lines were novel. Like the Porsche 356, the XK120 included many design features that would persist for decades in future Jaguars: a long hood, a voluptuous form, the absence of chrome trim, and a kidney-shaped grille. Unlike the Porsche, however, the Jaguar was not only a purposeful driving machine. Without being in any way a period piece, it unabashedly evoked prewar grand touring cars—that is, it was glamorous. The owner of the first American export model was the screen idol Clark Gable.

Over the next dozen years, the XK120 was followed by the XK140 and XK 150, which maintained the body design while improving mechanical performance and creature comforts. In

Jaguar XK120, 1948

the mid-1950s, work began on a successor, but William Lyons determined that the tooling costs would be too high and ordered that the company focus on sedans. Heynes, who was interested in sports cars, independently carried on developing the new model. The so-called XKE was based on an earlier Jaguar racing car, the XKD, whose body had been the work of Malcolm Sayer (1916–1970), an aircraft engineer who had joined Jaguar in 1950. Sayer's application of aerodynamic principles to racing car design had borne immediate fruit, first with a car that won Le Mans in 1951 and 1953 (the first British car to win since 1935), and then the D-type Jaguar, which took first place at Le Mans three years in a row. To reduce weight, this car had a semi-monocoque construction—a triangulated subframe carried the engine, transmission, and front suspension, and was attached to a monocoque shell that enclosed the cabin. This arrangement was repeated in the XKE, and Haynes added technical refinements such as disc brakes, rack-and-pinion steering, and independent front and rear suspension. The 3.5-liter XK engine with triple Weber carburetors produced 265 horsepower and an impressive top speed of 150 miles per hour. By now, Lyons had given his reluctant approval to the project, and when the XKE was unveiled at the 1961 Geneva Motor Show, much to his surprise, the car was an overnight sensation. The same happened at the New York Motor Show, which produced more than five thousand orders. *Car and Driver* described the car as "very fast, very stable, and, all in all, probably the car we'd most like to own of any we've tested in many a month."

At $5,600, the Jaguar XKE was hardly inexpensive—the same

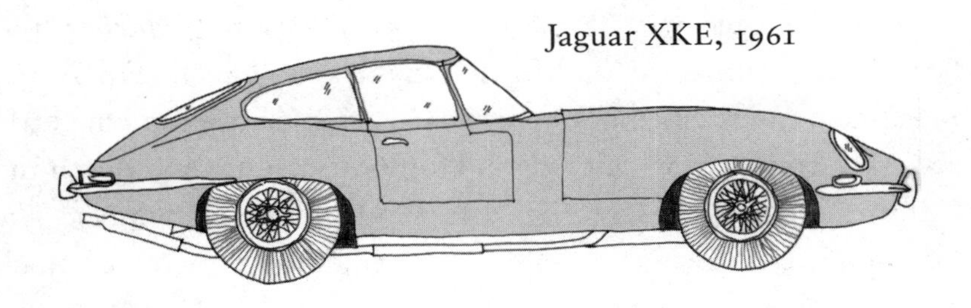
Jaguar XKE, 1961

price as a Porsche 356—but it was half the price of a comparable Aston Martin, Mercedes-Benz, or Ferrari, and it was considered a good value. In addition to exceptional performance and technological refinements, it was the car's appearance that made the strongest impression: a sculpted body, an extremely long hood, and a low, feline stance. Sayer had an unusual work method: using the chassis dimensions provided by the engineers, he calculated airflow and resistance using a slide rule and logarithmic tables, only later confirming the results in wind tunnel tests. In fact, despite its fluid shape, the XKE did not have a particularly low drag coefficient—only 0.5, compared to the Porsche 356's 0.36. That may be because Sayer, a watercolorist and a musician in his spare time, combined his aerodynamic analysis with a highly developed aesthetic sense. Perhaps the most remarkable aspect of Sayer's design was its durable appeal. In 2020, a *Road & Track* writer borrowed an old XKE for a day's drive. Almost sixty years had passed since the Jaguar's introduction, and the car had lost none of its appeal. "The great thing about an E-Type is how it brightens nearly everyone's day. No other car I've ever driven has received so much admiration, so many smiles and thumbs-ups . . . And all of this goodness is accompanied by a delightful, snarly straight-six soundtrack that'll make you wish we still used carburetors."

SPORTS CARS OFTEN FEATURED on the big screen in the sixties: Michael Caine in a Jaguar XKE in *The Italian Job*, Paul Newman in a beat-up Porsche 356 in the private-eye film *Harper*, and Sean Connery in an Aston Martin DB5 in *Goldfinger*. Perhaps the most memorable cinematic sports car of that decade was the little red Alfa Romeo in which Dustin Hoffman sped up and down the California coast in Mike Nichols's Academy Award winner, *The Graduate*. That two-seater—a Duetto Spider—attracted so much attention that Alfa Romeo eventually marketed a "Spider Graduate" version.*

The first cinematic starring appearance of an American sports car was the black Chevrolet Corvette that the actor Ralph Meeker drove in Robert Aldrich's 1955 film noir, Mickey Spillane's *Kiss Me Deadly*. This was many people's first view of a car that had been introduced only two years earlier. In 1952, seeing the success that imported sports cars such as the Jaguar XK120 were having in the United States, Harley Earl had his Styling Section produce a two-seater concept car. Despite Earl's predilection for extravagant forms and chrome trim, the design of the EX-122 was decidedly low-key, except for incipient tail fins that would disappear a few years later. One of the members of the design team was Henry de Ségur Lauve (1910–2017). Lauve, born in New Jersey but educated in Switzerland and France, had worked in Paris as a fashion designer and illustrator, and he brought a stylish, continental touch to the svelte EX-122.

The decision was made to go ahead with the project, and although GM's Cadillac division had argued strongly that the two-seater should be included in *its* lineup, the marketing plan dictated that the car should be moderately priced and the sports car was assigned to the entry-level Chevrolet division. The mechanical design was the responsibility of Edward Cole, who would work on the Corvair, and Maurice Olley (1889–1972), the British engineer and chassis expert who was head of research

* *Spider*, or *spyder*, of Italian derivation, is an alternate name for a two-seater roadster.

and development. To keep costs down, they used off-the-shelf Chevrolet parts for the chassis, an inline six-cylinder engine, and a two-speed automatic transmission. The skin was unorthodox, being made of fiberglass, a choice mandated by the complex body, which would have required expensive tooling had the material been steel. Fabrication in fiberglass was not only cheaper but faster, which allowed the car, now called the Corvette, to be included in the 1953 GM Motorama car show that debuted in the Waldorf Astoria Hotel in New York. The glamorous two-seater—white with a red interior—was a hit, and it arrived on the market several months later.

Chevrolet Corvette, 1953

Despite its auspicious launch, sales of the Corvette faltered. The $1,800 price announced at the Motorama show proved unrealistic, and the final price tag was $3,498, at a time when a Chevrolet Bel Air convertible, with proper roll-up windows and a tight-fitting top (which the Corvette lacked), could be had for $2,200. Nor, despite its appearance, was the Corvette particularly sporty, with a lackluster suspension and a six-cylinder engine that produced a top speed of 108 miles per hour. On top of this, the car only came with an automatic transmission. Two years later, Ford unveiled its two-seater Thunderbird, which was not only less expensive but had a powerful V8 engine that provided better acceleration and a top speed of 115 miles per hour.

In its first year, Ford sold 16,000 T-Birds; that same year, GM sold only 700 Corvettes. The GM sports car seemed destined for oblivion.

The Corvette's unlikely savior was a peripatetic Russian émigré, Zora Arkus-Duntov (1909–1996). Born in Brussels of Russian-Jewish parents, he grew up in St. Petersburg during the First World War and the Soviet revolution. In 1927, his stepfather, an electrical engineer, was reassigned to Berlin, where the young Arkus-Duntov studied mechanical engineering and began a life-long love affair with cars and car racing. By the outbreak of the Second World War, the family was living in Paris. Arkus-Duntov enlisted in the French air force, and when France fell, he managed to arrange the escape of his wife, brother, and parents via Portugal to New York. There, the enterprising Arkus-Duntov and his brother started an engineering company that successfully manufactured munitions and aircraft parts during the war.

After the war, Arkus-Duntov revived his interest in high-performance automobiles. A skilled engineer, he developed an overhead valve conversion kit that dramatically increased the output of the Ford flathead V8 engine, a favorite of hot-rodders. He bought a Talbot Grand Prix racing car, but failed to qualify at the Indianapolis 500. He got involved with Allard, a small British car company that specialized in hand-built racing cars, which gave him another opportunity to drive—at Watkins Glen in the United States and Le Mans in Europe. In none of these pursuits was he an overwhelming success, and in 1952—at the age of forty-three—he concluded that he simply did not have the financial resources to advance independently and decided to join one of the large American car makers. He approached Studebaker, Ford, Lincoln-Mercury, Chrysler, and General Motors, but his efforts did not produce any results—his background was simply too unusual for the mainstream car companies.

The following year, Arkus-Duntov saw the new Corvette at the Waldorf Astoria car show. He was impressed. "Mechanically, it stunk, with its six-cylinder engine and two-speed automatic transmission," he recalled later. "But visually, it was superb." He renewed his efforts at General Motors, and after an interview with Cole and Olley, in 1953 he was invited to join the Chevrolet division. The maverick engineer was hardly a model organization man, but after a rocky start, the handsome and charismatic Russian was soon making his mark. He worked on such mechanical problems as fuel injection, a technology then being developed to replace carburetors with a more effective method that sprayed—"injected"—an aerated fuel mixture into the engine. The in-house *General Motors Engineering Journal* described Arkus-Duntov as "a well-qualified technical man who has personal knowledge of the entire history of the sport and racing car history." He convinced GM to allow him to drive at Le Mans, first for Arden and then for Team Porsche, for which he piloted a 550 Spyder to a class victory.* He also achieved a winning time in the 1956 Pikes Peak International Hill Climb in a Chevy sedan fitted with Cole's new small-block V8 engine. That same year, Arkus-Duntov drove a stock Corvette fitted with the V8 engine and a four-barrel carburetor at Daytona Beach, clocking a speed of 150 miles per hour.

As GM's recognized expert on sports cars, Arkus-Duntov led a transformation of the Corvette, and by 1957 the car had a 4.4-liter V8 engine, a four-speed manual transmission, heavy-duty brakes and suspension, and an improved chassis. The fuel injection system he designed produced 283 horsepower and an impressive top speed of 132 miles per hour, which put the Corvette in the same class as a Jaguar. "Even the Anglophiles now readily admit that the Corvette will go. The only question left is how *well* it goes," observed *Road & Track*. After completing a road test, the magazine concluded that the performance of the 1957 Corvette was "unequalled by any other production sports car." The Corvette would go through many iterations, not least the striking 1963 Stingray, but for many people

* The 550 Spyder was the same car that actor James Dean was driving when he had his fatal 1955 highway accident.

that first-generation car, with Lauve's body and Arkus-Duntov's mechanical improvements, remains a classic.

THE MUSTANG WAS A different sort of fun car. It was the brainchild of Lee Iacocca when he was still at Ford. He had joined the Ford Motor Company in 1946, starting as an engineer, proceeding into sales, and rising to become general manager of the Ford Division. Encouraged by the early success of the Corvair Monza, his idea for the Mustang was a mainstream American sports car—not a competitor to the Corvette, but a reasonably priced compact that would cost less than $2,500 and be fun to drive.

Development of the Mustang started in 1962. This was only a few years after the Edsel debacle, so management kept a keen eye on expenses: the chassis was borrowed from the Ford Falcon, as was the engine and the three-speed manual transmission; other components came from the Fairlane and Galaxie. The use of standard parts meant that buyers could be offered many options: four engines, a manual or automatic transmission, different suspensions, convertible or hardtop bodies (and, later, a fastback), and various grades of interior finish. Standard in all the cars were two details that were novel in mainstream American cars of that period: bucket seats in the front, and a floor-mounted gearshift.

The styling team that worked on the Mustang was headed by Romanian-born Joe Oros (1916–2012), who was the director of Ford's Design Studio. Oros, who had studied automotive design in Sweden and was a graduate of the Cleveland Institute of Art, had worked for Harley Earl in the Cadillac division and later for the independent automotive stylist George W. Walker (1896–1993). In the 1950s, when Walker moved to Ford to head its new styling section, Oros accompanied him. One of Oros's first projects was the second-generation four-seater Thunderbird. The Mustang was likewise to be a two-door four-seater, but one that gave the impression of a roadster. The long hood and short trunk resembled the original T-Bird, and although the Mustang had a longer wheelbase, at 2,194 pounds it was considerably lighter. Purists might complain about the simulated side air intakes, but the overall impression was of a sophisticated, sporty vehicle. The car was unveiled in the Ford Pavilion at the 1964 New York World's Fair, and went on sale immediately. It proved a runaway success, selling 550,000 in the first year and one million in less than two years.

Ford Mustang, 1964

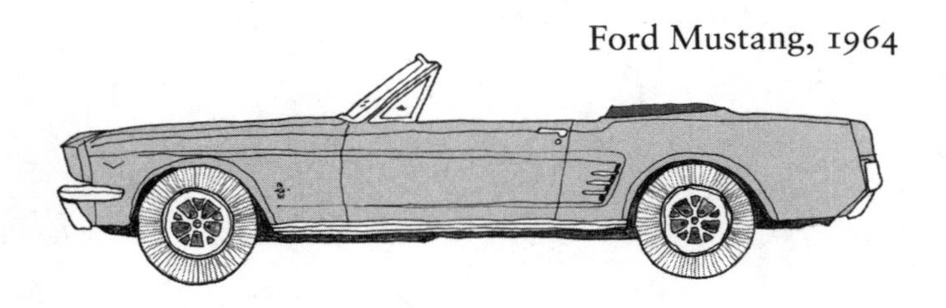

The Mustang attracted a lot of buyers—Oros had been instructed that his design should appeal to women as well as men. The base model came with a 2.8-liter six-cylinder engine that produced 101 horsepower and a top speed of 96 miles per hour, more than enough for most car owners. Responding to demands from enthusiasts for larger engines, Ford offered optional V8s of increasing size. The racing driver and automotive designer Carroll Shelby (1923–2012), who was responsible for the Ford GT40, produced high-performance versions of the Mustang suitable for road racing. That was the car Steve McQueen drove in a hair-raising San Francisco chase scene in Peter Yates's *Bullitt*. The dark green 1968 GT Fastback had a 6.4-liter V8 engine that produced 325 horsepower and a top speed of 122 miles per hour.

The Mustang came to define a new category that people called a "pony car," referring to the horse on the car's grille emblem. A pony car was not "a vast boudoir on wheels," in Maurice Olley's memorable phrase, but a compact coupe with a long hood and a short deck, or trunk, a car that could carry four people and was both sporty and affordable. And fun.

CHAPTER
NINE

MADE IN JAPAN

If you make a superior product, people will buy it.

—SOICHIRO HONDA

THE CLOSEST I CAME TO OWNING A SPORTS CAR WAS A Honda Prelude. Prelude was a not-very-serious name (prelude to what?) for a serious car. The front wheels were driven by a 1.8-liter four-cylinder engine that put out 100 horsepower. Because the Prelude was relatively light, this made for a top speed of 112 miles per hour, and a zero-to-sixty acceleration of nine seconds. The four-wheel double-wishbone independent suspension, which used two arms to locate the wheel, as well as front and rear stabilizer bars, produced excellent handling. *Road & Track* called the 1983 Prelude "the best car in its class . . . a benchmark car that will send competitors back to the drawing boards."

The Prelude was what the industry called a sports coupe, or a 2+2—a two-door two-seater with an auxiliary rear seat just large enough for a small adult or two children. The car had debuted in 1978, and my four-year-old navy blue Prelude was from the second generation that appeared in 1983, with a slightly longer wheelbase, more powerful engine, and redesigned body. The notchback form, popularized by the Mustang, was largely the work of Hiroshi Kizawa (1936–2016), the company's chief

Honda Prelude, 1983

designer. He gave the car a low hood line, with concealed pop-up headlights that contributed to the aerodynamic design and a low drag coefficient of 0.36.

The Prelude really was a benchmark car. The second generation offered antilock braking, which prevented the wheels from locking up during hard braking, giving the driver greater control of the car. Experiments with antilock brakes dated back to the 1920s, and the first mass-produced car with electronic antilock brakes on all four wheels was a 1960s Mercedes, which used an expensive system developed by Bosch. Honda's version, introduced in the Prelude, was a more affordable option, and in some European countries it was offered as standard equipment. An even more revolutionary feature that was an option in the 1987 Prelude was four-wheel steering, which was said to improve handling as well as parking maneuverability.

My Prelude didn't have antilock brakes, and four-wheel steering was still in the future, but such technical refinements aside, the great pleasure of the car was in the driving. Seated low in the bolstered seat, grasping the thick steering wheel, with the moon roof open, I could whip along a twisty back road on my way to the village to do errands—playing at being Juan Manuel Fangio in the Mille Miglia. As with the Mini Cooper, it wasn't the speed but the fact that it had great "road feel." According to *Road &*

Track, "Honda engineers told us they used the Porsche 924 as the handling model for the Prelude and, as far as we're concerned, they hit it dead center."* At the same time, the nimble Prelude was an accomplished touring car. We once drove comfortably from our home in southern Quebec to Camden, Maine—a six-hour trip—with an elderly friend in the rear seat. I remember long trips in the Citroën Deux Chevaux and the Mini, temperamental cars that required constant attention and fiddling—something often seemed to go wrong. Nothing ever went wrong with the reliable Prelude. But of course, it was made in Japan.

I am old enough to remember when "Made in Japan" meant cheap ballpoint pens and poorly made consumer products, but by the late 1960s that had changed. Brands such as Sony, Nikon, and not least Honda, signified reliability, durability, and design quality. It was a remarkable transformation. In 1987, when I bought the Prelude, the Honda Motor Company had been making cars for only two decades. Soichiro Honda (1906–1991), the son of a village blacksmith, was a self-taught car mechanic who, before the war, ran a small company supplying piston rings to Toyota. In 1946, together with the businessman Takeo Fujisawa (1910–1988), he founded a company to manufacture motorized bicycles using army surplus generator motors. Eventually, Honda moved up to mopeds and scooters. In 1958, he produced the Super Cub, a lightweight motorcycle with a four-stroke single-cylinder 50 cc engine that revolutionized the industry. Other engineering innovations followed—electrical starters, disc brakes, larger engines—that added to Honda's reputation as a first-class motorcycle maker that could compete with the established German and Italian marques.

By the mid-1960s, Honda was the world's largest manufacturer of motorcycles, and ready to expand into automobiles. The

* The 924, an entry-level 2+2 sports coupe introduced in 1976, was the first Porsche with a water-cooled front engine driving the rear wheels. In 1983, the 924 sold for about $16,000, compared to $9,000 for a Prelude.

first mass-produced model was the Honda N360, a minicar in the *keijidōsha* (light automobile) class, the smallest highway-legal passenger car. The Japanese government had introduced the "kei car" category in 1949 to encourage the manufacture of small, affordable cars that were suited to the postwar economy. These cars were popular not only because of their low prices but because they offered tax and insurance advantages—even a simpler driving test. The kei standards limited body dimensions and engine capacity, and were periodically upgraded so that, by the 1960s, the maximum allowable dimensions were slightly less than ten feet long and about four feet wide; the maximum engine displacement was 360 cc, the equivalent of a small motorcycle.

Soichiro Honda thought that existing kei cars were dangerously underpowered, and the N360 was given a high-revving motorcycle-type engine that produced a top speed of sixty-five miles per hour. The transversely mounted two-cylinder 354 cc aluminum engine, which was air-cooled and drove the front wheels, was so compact that there was room under the hood for the spare tire. The two-door monocoque body was the work of a team working under Motoo Nakajima. Like the 1959 Mini, which it resembled, the N360 had twelve-inch wheels that were pushed out to the corners to improve handling. Both front and rear wheels had independent double-wishbone suspension. To reduce weight, the rear hatch and the dash were fiberglass. The rear seat folded flat to increase cargo space.

In 1969, Honda decided to export the N360 to the world's leading car market, the United States. The renamed N600 had an identical body with a larger 599 cc engine that produced a top speed of eighty miles per hour, which was considered better suited to American highway driving. Honda sold more than 25,000 cars over the next three years, until American emission standards, which had exempted cars below 850 cc from controls, were tightened. But by then, Honda had a successor ready.

The revolutionary Honda Civic was a hatchback with a front engine and front-wheel drive. While the body of the N600 had

Honda N360, 1967

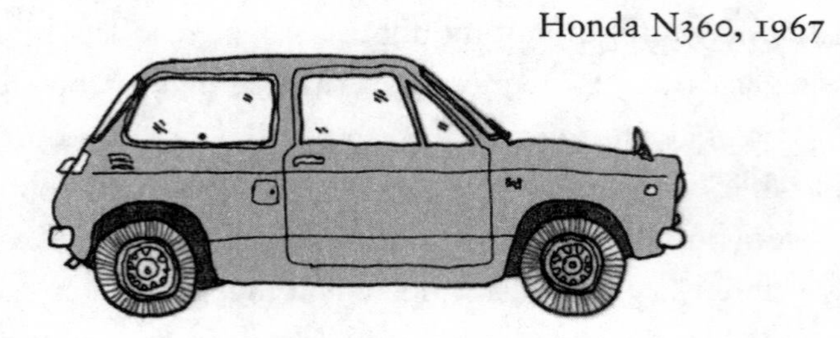

Honda Civic, 1972

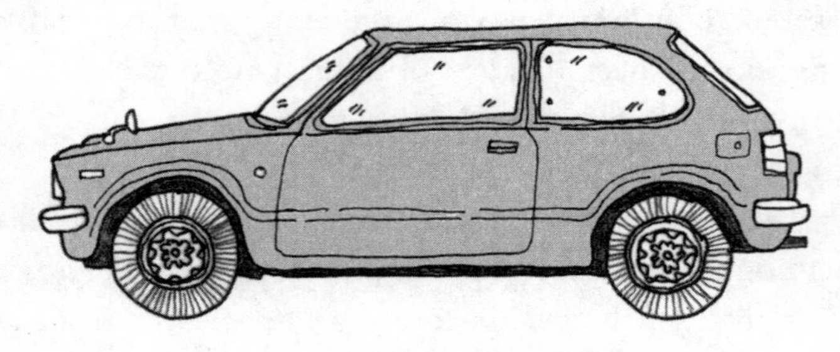

been modeled on the Mini, the Civic design was original. The wheelbase approached twelve feet, making it larger than a microcar, although smaller than most compacts of that era. Soichiro Honda, who had been stubbornly committed to high-revving air-cooled engines, despite their disadvantages of being noisy and having a shorter life than conventional car engines, had finally relented and his engineers developed a 1.2-liter four-cylinder water-cooled engine putting out fifty-two horsepower and a top speed of eighty miles per hour. With front and rear independent suspension, the car handled well; brakes were power-assisted—discs in the front and drums in the rear.

To maximize comfort, the Civic was designed from the inside out. The longer wheelbase provided an exceptionally roomy interior for such a small car, and the folding rear seat gave ample cargo space. The straightforward dashboard design was unusually spare for its time. The design of the body, by Shinya Iwakura and

Hiroshi Kizawa, was more like a functional wrapper than styling. The following year, Honda introduced its CVCC (compound vortex controlled combustion) engine, which produced more efficient combustion without expensive fuel injection and enabled the Civic to meet Japanese and American emission standards without a catalytic converter, hence not requiring (more expensive) unleaded gas. "Gasoline made simple," boasted the Honda ad.

The Civic could not have debuted in the United States at a more auspicious time: the following year saw the 1973 oil crisis, when the international oil cartel OPEC embargoed oil delivery to nations that had backed Israel in the Yom Kippur War. The result was widespread gas shortages, rationing, and long lines at service stations. In addition, prices rose, and by the end of the year the price of gas at the pump had doubled. The oil crisis opened the door to what is sometimes called the "Japanese invasion." American buyers discovered that Japanese cars were not only fuel-efficient but also well built and reliable. They were cheaper, too, because unlike American manufacturers, the Japanese limited the number of buyer options, thus reducing production costs. In 1974, the Environmental Protection Agency conducted a fuel economy test of 376 vehicles; the fifty cars with the best mileage were all imports, the top three were Japanese, and the car with the best fuel economy was the little Honda Civic.

Annual sales of the Civic, which started slowly with 32,000, grew quickly, surpassing one million in 1981, two million in 1988, and three million in 1993. Worldwide, the Civic became Honda's best-selling model. The appearance and size changed over the years because, unlike the Volkswagen Beetle and the Renault 4, the Honda Civic was not a specific design but an idea: a small, well-engineered, and practical hatchback that was not only affordable but also fun to drive. As I write, the Civic is in its eleventh generation. Truly a global car, it is manufactured in the United States, the United Kingdom, and South Africa, as well as in Japan. Despite having grown into a full-fledged compact, it is actually lighter than the 1972 original. Civics are now available with a turbocharged engine, as a sporty performance version, as well as a hybrid. *Car and Driver*, unable to choose between them, named the entire Civic lineup in its Ten Best Cars for 2023.

There was a second Honda on the *Car and Driver* Ten Best list: the Accord.* Honda introduced the Accord in 1976. It was a compact front-engine, front-drive hatchback—in effect, an enlarged Civic, but with a more powerful engine, a quieter ride, more refined suspension, and power steering. A year later, the hatchback was followed by a four-door sedan. The lightweight vehicle was a roomy four-seater with excellent handling and impressive fuel economy—better than forty miles per gallon on the highway. The manual transmission had five speeds instead of four, and an automatic transmission was an option. The car's standard features included cloth seats, an AM/FM radio, a tachometer, day/night adjustable rearview mirror, and intermittent wipers. This added up to an exceptionally well-equipped car selling for less than $4,000.

Honda Accord, 1990

As the Accord sedan grew in popularity, it also grew in size, and by the time the fourth-generation model was released in 1990, it was competing successfully in the popular midsize family-car category. In 1991, the Accord became the best-selling car in the United States, the first import to hold that position

* This was the thirty-seventh time that the Honda Accord appeared on *Car and Driver*'s Ten Best list. An unrivaled record.

(although by then the car was manufactured in Ohio). The 1990 Accord demonstrates how the design of a mainstream four-door family sedan had evolved over the years. The car had a low drag coefficient of 0.33—the same as the Citroën DS, but unlike the French car, the aerodynamic styling was inconspicuous. The 1990 Accord was the first car to use clear-lens multi-reflector headlamps in which the molded plastic reflector, rather than the lamp, directed the light beam. The result was that the headlamps were no longer prominent "eyes" but were smoothly integrated into the body shape. The sleek body had no fenders, no chrome bumpers—in fact, no chrome at all. The prominent front grille of yesteryear was gone, too, replaced by discreet slots; the sole side-body trim pieces were sensible rubber scratch protectors. The remaining traditional feature was the rearview mirror, a device that originated in a 1911 Indianapolis racing car, but did not become standard in production cars until the mid-1960s. In all, it was as if Harley Earl's playbook had been tossed aside, with the exception of his enduring "longer and lower" dictum; the Accord approached sixteen feet in length and was only four feet, six inches high.

Soichiro Honda retired from active management in 1973, but in many ways the 1990 Accord reflected his emphasis on technical innovation and performance. This focus was particularly visible in another Honda model that was introduced that same year, a two-seater sports car. The Honda NSX was not a sports coupe like the Prelude, but an exotic high-performance roadster. Why would a manufacturer of mainstream automobiles build such a car? Motorsport was a Honda tradition. Like many auto pioneers, the young Soichiro Honda was an avid racer, at least until he was seriously injured driving a supercharged Ford. Beginning in 1959, Honda factory teams competed in Grand Prix motorcycle races, and in 1964 Honda built and raced its own Formula One racing cars, later entering modified Civics and Accords in rallies. In the 1980s, Honda supplied engines to British racing teams, and for several years the McLaren-Honda Formula One car dominated its category. Such participation served to establish Honda as a serious high-performance manufacturer, so it was perhaps inevitable that the company would enter the exotic sports car category. Honda's goal was to use its racing experience to build a production car that matched the performance of a hand-built Italian sports car while offering everyday drivability and reliability, all at a reduced price—under $60,000. The ambitious benchmark was the Ferrari 328, a very fast (160 miles per hour) mid-engine roadster that looked like it belonged at Le Mans.

The mid-engine NSX had rear-wheel drive. The engine was a V6, compared to a V8 in the Ferrari, but the NSX was lighter and its all-aluminum three-liter engine produced a similar top speed. The aerodynamic body had a drag coefficient of 0.32, and at such high speeds, the problem was not only wind resistance—the body design also had to create a downforce on the rear wheels to maintain tire grip on the road. Shigeru Uehara (b. 1947) headed the engineering team, and the styling team was led by Masahito Nakano (b. 1957). The NSX was slightly shorter than the Accord, but wider and lower. The stiff stressed-sheet aluminum body was reinforced by a lattice of forged alloys. The engine location, behind the driver, required prominent side air-cooling scoops. Otherwise, the smooth design, with its sloping front and concealed headlights, was a model of Japanese understatement.

The NSX was well received by the automotive press. According to the British motoring journalist L. J. K. Setright, the NSX was "sublimely superior to everything . . . what the NSX really amounted to was the world's fastest, safest, and most beautifully made luxury car." In September 1990, *Car and Driver* road-tested five high-performance two-seaters—a Porsche 911, a Corvette ZR-1, a Ferrari 348, a Lotus Esprit, and an NSX—and gave the Honda first place. The magazine wrote that the NSX excelled in all categories: engine, handling, controls, build, and comfort. "Honda revitalized the motorcycle market in the 1960s and reshaped the small-car market in the '70s. Now it's going to teach the world how to build Learjets for the ground." It didn't turn out that way. The NSX did not sell well, despite its sterling qualities and its attractive price.* The car was marketed in the United States under the Acura badge, which was Honda's luxury division, founded four years earlier, but the Acura name simply

Honda NSX, 1990

Mazda MX-5 Miata, 1989

did not have the cachet of Ferrari and Porsche. American sales fell to a few hundred a year, and Honda discontinued the car in 2005.

While the NSX fared poorly, a very different Japanese two-seater was a resounding success. The Mazda MX-5 Miata was a conscious reinvention of the little British roadsters of the 1960s, such as MGB and Triumph—weekday transportation that could be driven on the racetrack on weekends. "We wanted to combine the reliability and quality of a Japanese car with the excitement and emotion of an inexpensive, lightweight, rear-drive convertible," said Bob Hall (b. 1953), the American automotive journalist who suggested the concept to Mazda and who was later

* In 1990, the NSX sold for $58,000, compared to $77,000 for the Porsche 911 and $101,000 for the Ferrari 348.

hired by the company to manage the project. The Miata was an enthusiast's vehicle, with a relatively small engine (1.6 liters, four cylinders, 116 horsepower), five-speed manual transmission, and a compact open-air cockpit. The light car, with front engine and rear drive, handled excellently. The body design, which came from Mazda's Southern California studio, recalled the earlier Lotus Elan, a classic of the 1960s. But the Miata was not a retro-style car, it was a modern Japanese automobile—that is, unlike its British predecessors, it was well built and dependable. On top of all that, it was affordable, costing a fraction of a German or Italian roadster. "It works. Its top goes up and down easily. It's cute. It's responsive. Girls love it. Men love it. Everybody loves it," wrote *Car and Driver*, which included it on its Ten Best list the following year.

I HAD NEVER OWNED a Miata—my friend Ralph once gave me a ride in his and I found the little sports car a tight fit. But over the years, Shirley and I did have several Japanese cars in addition to the fun Prelude: a bulletproof Toyota Celica that I never warmed to, an Infiniti (more about which later), and a Subaru station wagon (two of them, at different times), the Japanese car we owned the longest. The Subaru was not an especially pretty car—an unremarkable, boxy design with no particular flair. "While this might sound gratuitous, it could be said that 'Subaru' and 'styling' are mutually exclusive terms," wrote Brock Yates in *Car and Driver*. "One sometimes wonders if in fact there is a Subaru styling department, or if there is merely a blackboard mounted on a sidewalk outside the factory where passersby can scribble suggestions." Things were no better in the interior. "The Subaru insides are relieved only by inserts of polyester cloth in the seat bottoms," Yates wrote. "The instrument panel is an ergonomic comedy, with all manner of buttons and knobs, plus digital readouts that emit a red-and-green night glow that resembles a Kmart Christmas display." This was not an exaggeration. The

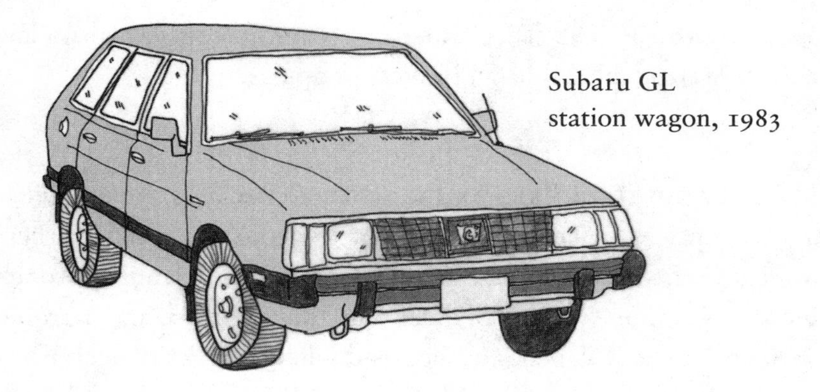

Subaru GL
station wagon, 1983

first time Shirley and I drove our new 1983 Subaru in the dark, the dash lit up exactly like a Christmas tree.

Like most Subaru owners, we didn't buy our car for its looks, but rather for a single—and singular—feature: four-wheel drive. In 1972, Subaru was the first mass-market manufacturer to offer four-wheel drive in a station wagon or sedan.* Moreover, Subaru's four-wheel-drive option was affordable—a $500 extra on our 1983 wagon. No wonder the four-wheel-drive Subaru became a bestseller in snowy places such as Colorado, Vermont, and eastern Canada, where we lived.

"Inexpensive, and built to stay that way" was Subaru's motto. The tried and true engine was a flat four, the same venerable boxer that had been used years earlier in the KdF-Wagen, producing a distinctive low-pitched rumble from the exhaust. The Subaru engineers had teased sixty-seven horsepower out of the 1.6-liter engine, and, like the Civic, the Subaru did not require a catalytic converter or unleaded gas. The four-wheel drive was engaged simply by shifting a lever. The passenger compartment may have been all plastic, but despite the car's small dimensions—the ninety-seven-inch wheelbase was considerably shorter than an Accord—the tall interior was roomy and provided generous cargo

* Before the Subaru, four-wheel drive was available only on all-terrain vehicles such as Jeeps and Land Rovers. The lone exception was the 1966 British Jensen FF, a limited-production (only 320 were built) luxury touring car.

space. With the rear seats folded, we transported skis and bicycles, as well as many bushel baskets of apples.

EUROPEAN CAR BUYERS IN the 1980s appreciated the reliability and low prices of Japanese cars, but they did not consider them to be serious competitors to established national brands. Europe's motoring heritage was much older than that of Japan, and marques such as Morris, Citroën, and Fiat had a long history of such pace-setting designs as the Mini, the DS, and the Cinquecento. Just as Parisian couturiers were global leaders in women's fashions, Italians and Germans maintained a special position in car design. Even the British, who had lost their leadership in the motoring field, were still capable of producing a stylish car like the Range Rover. To immerse their staff in this design culture, the leading Japanese car makers established design studios in Europe: Toyota in Brussels, the capital of the European Union, and Honda and Nissan in Germany, Europe's largest car market.

One of the first products of Nissan Design Europe, which was located in Munich, was the Primera, a midsize sedan launched in 1990. The Nissan engineers combined front-wheel drive with a multi-link suspension—no one had done this before—which produced the sort of exceptionally agile handling that European drivers appreciated. The two-liter, four-cylinder, sixteen-valve engine with fuel injection was high-revving and put out a respectable 115 horsepower, which powered the light car to a top speed of 124 miles per hour. Europeans valued braking and safety as well as handling, and the Primera delivered on that score, too. Reviewers of the car were surprised when they drove it because the somewhat bland styling did not promise outstanding performance. The conservative design was European in the sense that it could easily be mistaken for a Peugeot 406 or an Opel Vectra—hardly striking cars—and the best that most reviewers could say of the low-key Primera body was that it was inoffensive.

Infiniti G2o, 1995

The Primera, rebadged as G2o, served as the entry-level model for Infiniti, Nissan's newly founded luxury division. "Born in Japan. Educated in Europe. Now Available in America" was how Infiniti advertised the car. The G2o was basically a top-of-the-line Primera, with power everything, air-conditioning, four-wheel disc brakes, a five-speed manual or automatic transmission, antilock brakes, dual airbags, a leather interior, and a Bose sound system. To top it off, unlike the Primera, which was manufactured in England, the G2o was made in Japan.

The British motoring press sometimes referred to the Primera as a "mini BMW," and I came across a review that described the G2o as an updated BMW 1600. That piqued my interest. It had been more than a decade since I had driven what was one of my favorite cars, and since I occasionally passed by an Infiniti dealership, I thought I should stop and take a look. I was the only customer in the showroom. Although the big Infiniti Q45 sedan had drawn raves from reviewers, the marque had been largely overshadowed by Lexus, which had captured luxury car buyers' imaginations. A salesman hurried over and was happy to let me take a test drive. The exterior may have been unprepossessing, but even a short time behind the wheel convinced me that the comparison to the BMW was not specious. The G2o was responsive and fast. In addition, the car was exquisitely built, and the attractive interior included an analog clock, just like my old 1600. I was touched by the trivial detail.

I had the impression that G20s were not exactly flying off the shelf, and the salesman was anxious to make a deal, offering a sizable discount as well as taking my beat-up Subaru wagon as a trade-in. I returned a few days later with Shirley for a second test drive. We had recently moved to suburban Philadelphia, and I pointed out that we really didn't need a four-wheel-drive wagon anymore, especially one that was starting to show signs of delayed Canadian rusting. While generally skeptical of my motoring enthusiasms, she had to admit that the little Infiniti was a very nice car. I returned a few days later to sign the papers and take delivery.

My black 1995 G20 lived up to expectations. It was a pleasure to drive. Unlike our old Subaru, it cosseted us in luxury; moreover, it was air-conditioned, which was vital for Philadelphia's hot and humid summers. The G20 was entirely reliable, delicate looking but solidly built. It had only one failing, which I discovered after my first daylong drive. After a few hours, I started to fidget. It wasn't the seat, which was fine, but rather the legroom—there wasn't quite enough. When I checked online data, I found that on paper, the legroom of my Subaru was actually an inch shorter, yet I had never felt constrained. Perhaps it had to do with the driving position, which had been more upright than in the G20. Shirley, always the realist, did not beat about the bush. "You really should have a larger car," she said. Or did she say "deserve"? No, that was probably my imagination.

CHAPTER
TEN

END OF THE ROAD

Oh Lord, won't you buy me a Mercedes-Benz?

—JANIS JOPLIN

THE 1990S HAVE BEEN REFERRED TO AS THE JELLY BEAN Era in car design. Gone were the boxy shapes and creased edges of the previous decades; instead, cars were styled with smooth shapes, rounded corners, and integrated or pop-up headlights. This was largely the result of a focus on increasing aerodynamic performance in order to improve fuel efficiency. By now, computer modeling had augmented wind tunnel testing, which greatly facilitated the design process and led to a certain homogenization among manufacturers. This was intensified by the growing popularity of SUVs, whose two-box configuration did not leave much room for variation—there was little apparent difference between an economical Kia and a luxury Lexus, except that the Lexus had a larger—and to my eye, unsightly—grille. The 1990s did see some unusual cars, such as the burly Hummer, the civilian version of the US military's Humvee—a replacement for the jeep—and the long-nosed Dodge Viper, an extremely powerful roadster that *Car and Driver* described as "one of the most exciting rides since Ben Hur discovered the chariot."

Neither the Hummer nor the Viper was the car for me, and while I was in no hurry to replace our Infiniti, I occasionally stopped at dealerships to look at cars and take a test drive. I fondly remembered Jean-Louis's old Saab, but that company was now half-owned by General Motors and the car had lost much of its offbeat charm. I tried a Lexus, but although it was roomy and well built, I found its bland sense of luxury unappealing. Shirley and I occasionally shopped at a garden supply center on Philadelphia's Main Line, which was near a Mercedes dealership. A German car had been my introduction to motoring, and ever since my decade with the BMW 1600 and a brief dalliance with a secondhand Audi 4000, I had admired the no-nonsense approach of German car makers. I remembered once being given a ride in a Mercedes by a Montreal friend who worked in the university's medical research lab. She had never shown much interest in cars, but she told me, "This is the first car I've driven that's as well designed as my electron microscope."

Shirley and I visited the Mercedes dealership. We couldn't afford a new car, but several used—"pre-owned"—sedans were parked in the rear lot. We took turns driving. "It feels like a truck," said Shirley. She meant it approvingly; despite its agility, the car—larger than the Infiniti—had a truck-like solidity. Most of the cars on the lot had leather interiors, leather being de rigueur in upscale cars. Do you have any with vinyl? I asked the salesman. The Germans had perfected long-lasting vinyl upholstery—we had had it in the BMW, and I remembered it as pleasant to sit on, easy to clean, and dog-proof. The dealer had one car finished in perforated MB-Tex, which is what Mercedes called vinyl. The four-door sedan was dark maroon—"Barolo Red"—with a creamy-beige interior. And plenty of legroom.

The three-year-old Mercedes—a 1993 model—belonged to the W124 class, a midsized car that Daimler-Benz had introduced almost ten years earlier. The W124 included such safety features as four-wheel disc brakes with antilock control, dual airbags, and front and rear crumple zones. The concept of crumple zones protecting a rigid passenger compartment was the brainchild of Béla Barényi, the Hungarian engineer whose 1920s sketches for a "people's car of the future" had anticipated the KdF-Wagen.

Mercedes-Benz
W124 300E, 1993

Barényi joined Daimler-Benz in 1939, and starting in the 1950s, the company incorporated many of his safety features in its cars.

The W124 included many other small but useful innovations. The spacious trunk opened almost to the bumper to facilitate loading, a feature later adopted by other car makers. The height, tilt, and headrest of the motorized front seats were adjusted by means of an intuitive control that resembled a miniature car seat. The rear headrests could be remotely dropped out of the way by the driver to increase rear visibility, although they had to be raised manually; the rear window shelf contained a built-in box with a first aid kit. The single windshield wiper had an eccentric action whose sweep was larger than other cars' dual wipers. The headlights, which included fog lamps, had their own mini wiper-washers that were automatically triggered whenever the headlights were on and the windshield washer was activated. A convenient mini visor shaded the small leftover space over the rearview mirror when the two sun visors were lowered. Another telling detail: the exterior rearview mirrors were asymmetrical: the driver's-side mirror was rectangular, but the passenger side was square, to give the driver a deeper field of view.

The model of W124 we were looking was a 300E, the *E* standing for *Einspritzmotor*, or fuel-injected engine, whose use dated back to 1954, when Daimler-Benz was the first manufacturer to use this technology in a mass-produced car. The 300E was an engineer's idea of a luxury car—no plush, no frivolity, a severe interior compared to the Infiniti. Csaba Csere, in a *Car and Driver* review, called the 300E a "transportation tool."

Mercedes-Benz W124 300E, 1993

Since efficient transportation requires speed, the 300E was given the ability to go very fast. Since roads come in all combinations of straight, curved, rough, and smooth, the 300E has a suspension that can cope very capably with any road at high speeds. Since transportation often requires several hours, the transportation was made spacious and comfortable for its occupants. Since there is a failure to transport if the occupants don't survive the trip, the 300E was equipped with excellent active and passive safety features. Since a car can't convey anyone if it doesn't run, the 300E was built solidly and from the most durable materials. And since fuel is often scarce and expensive, the 300E was engineered to do all of the above relatively economically.

What Csere, an MIT-educated engineer whose car reviews tended to focus on technical detail, didn't mention was that the 300E was a very handsome tool. Unlike many cars of the 1980s, the body had no trendy creases or sharp angles, and its unusually narrow and rather tall body had rounded corners like a used bar of soap, although it was no jelly bean. The result was an impressively low drag coefficient of 0.28, lower than the 0.30 of our Infiniti. Over its production life, the 300E came with different motor options. The one we test-drove had a 2.8-liter six-cylinder engine with dual overhead camshafts, which improved the air-fuel mixture and achieved greater efficiency and better fuel economy. The engine put out an impressive 193 horsepower, more than many American V8s, and powered the 3,200-pound Mercedes to a top speed of 140 miles per hour, making it one of the fastest four-door sedans of that period.

Not that I was planning to drive that fast. I hadn't researched the 300E before we visited the dealer, nor had I read Csere's review. In truth, my car purchases were often impulsive—I either liked a car or not. And I liked this one. Shirley agreed, but she had one condition: we had owned ten cars during our twenty-three years together, and I had to promise that this would be the last one. We bought the car that same day, trading in the almost-new Infiniti, which turned out to be worth much more than I—or the salesman—anticipated.

A FEW WEEKS LATER, still caught up in the enthusiasm of new ownership, I was washing and waxing the car when I came across a curious detail. Instead of the usual rubber molding on the outside of the doors to protect against accidental scratches, the entire lower door panels were . . . plastic. The panels were the same color as the rest of the car, which is why I hadn't noticed them earlier. Snap-in plastic sideguards had been introduced by Mercedes in its flagship S-class in 1979 and had become part of the W124 class in 1989. German buyers, who were initially ambivalent about having plastic on a Mercedes, called them *Saccobrettern*, or "Sacco boards," after their designer.*

The Italian-born Bruno Sacco (b. 1933) was a mechanical engineering graduate of the Polytechnic University of Turin. He had been briefly employed by Ghia and Pinin Farina in Turin, but his ambition was to work for Mercedes-Benz, and in 1958 he joined the company's bodywork department, whose plant was located

* Plastic sideguards, which would become popular on many cars, were the idea of Renault designer Michel Boué, and they debuted in the 1972 best-selling Renault 5, a Mini-like hatchback marketed in North America as Le Car.

in Sindelfingen, near Stuttgart. Daimler traditionally considered body design an engineering problem and had only recently hired Paul Bracq (b. 1933), a French designer, to head its new styling section. Rather than playing second fiddle to Bracq, Sacco moved to the experimental and safety innovation department, where he spent four years under the legendary Barényi. When Bracq left to return to France, Sacco was appointed to lead a new design department that built operational experimental prototypes, including racing cars, microcars, and even a three-wheeled two-seater. In 1975, after being made the company's chief engineer, he became head of what was now a full-fledged styling department with a staff of 130, which he led until his retirement in 1999.

Sacco was responsible for some of the classic Mercedes, including the S123 wagon and the SLK roadster, as well as the W124 sedan. While he had long experience in performance and safety innovation, he considered body design to be an important ingredient in the success of a car. He summarized his philosophy in two rules. The first was, "A Mercedes-Benz needs to look like a Mercedes-Benz." What did that mean? American car design was influenced by the practice of introducing new models every year, which was intended to entice owners to trade in their old cars. Mercedes did not change its designs annually: the W124 would have a thirteen-year run; its predecessor, the W123, had remained in production with minor changes for nine years. When a new model was introduced, it generally reflected advances in engineering, safety, and performance, but when it came to styling, Mercedes tended to adopt a conservative approach that could be described as evolutionary rather than radically innovative.

Our slab-sided 300E, for example, was a direct descendant of the Mercedes 180, a midsized sedan that Daimler-Benz had introduced in 1953. The 180 was the company's first entry-level model of the postwar period. Intended for a broader market than its flagship S-class, the 180 was popularly known as the Mittleren Mercedes-Klasse, or the Middle Mercedes Class, and was as popular with taxi drivers as with midlevel business executives. The car was a

Mercedes-Benz 180,
1953

radical departure from prewar designs, doing away with traditional running boards and freestanding fenders, and integrating the headlights into a smooth body that Germans called *ponton* styling. The 180 was also the first Mercedes with a monocoque body, and the first to incorporate Barényi's front and rear crumple zones. The sober three-box shape and upright stance of this car were distinctly American compared to the streamlined Citroën and Jaguar sedans of that same period. Some features carried over from previous Mercedes models: the distinctive chrome-framed protective radiator grille, which originated in 1931, and the three-pointed-star hood ornament, which dated from the 1920s. Like all the Mercedes of the immediate postwar decades, the Mercedes 180 was designed by engineers. Fritz Nallinger (1898–1984), a graduate of the Technical University of Stuttgart and the technical director of Daimler-Benz, was head of the team, and the body design was the work of Karl Wilfert (1907–1976), a Viennese engineer who would oversee the famous 300SL gull-winged coupe, and who, as head of the bodywork department, would later hire Bruno Sacco.

The second Sacco styling rule was, "The next model should never make the previous model look old." Daimler-Benz's marketing strategy was to sell overengineered, durable, and expensive automobiles that would last—many Mercedes owners expected to keep their cars for at least two decades—so the styling had to last, too. That meant avoiding trendy features that would quickly become stale. For example, the W124, which had the same wheelbase as its predecessor W123, included many improvements: it was several hundred pounds lighter and slightly narrower, with a taller rear end, and it was also more aerodynamic, despite having a roomier interior. On the other hand, the boxy shape and the grille and hood ornament were carried over from the previous model, as were the pull-to-open door handles and the unmistakable ribbed taillights.* Thus, the two cars looked like members of the same family—second cousins, perhaps.

Mercedes-Benz W123, 1976

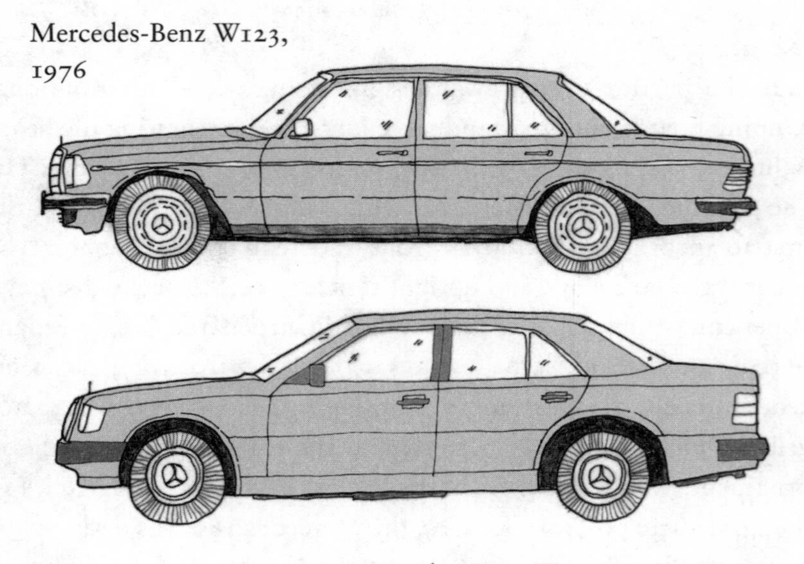

Mercedes-Benz W124 300E, 1984

Referring to the W124, Bruno Sacco said, "I believe that each and every Mercedes model has a certain potential to become a classic later." What is a classic? Car collectors use the term to indicate a car's age: "classics" are more than twenty years old, "antiques" are more than forty-five years old, and "vintage" automobiles are built before 1930. But age alone hardly qualifies a car as a clas-

* The ribbed taillights, designed by Béla Barényi, were intended to repel dirt and remain visible on muddy roads.

sic in the aesthetic sense. Nor does sales volume. The best-selling Volkswagen Beetle and the Renault 4 are classics not because they had long production lives, but because they raised automobile design to a new level. So did cars such as the VW bus and the Ford Mustang, which opened the door to new motoring categories. Then there are classics such as the Mini and the Porsche 356, whose designers solved a particular problem, or set of problems, with such panache that the cars are paragons of which it can truly be said that "nothing can be added or taken away." Cars such as the Land Rover and the quirky Citroën Deux Chevaux exemplify a distinctly original and single-minded approach to utility that qualifies them as classics. As for the Citroën DS, the Jaguar XKE, and the original Corvette, mechanical innovation aside, they are simply ageless beauties—Venus de Milos.

No one would call our boxy Mercedes a beauty. Its designers conscientiously solved many problems—of utility, safety, and performance—and many of the car's features would influence other car makers. In all, the 300E could be said to represent the triumph of good genes, and of ideas that had been percolating at Daimler-Benz since before the Second World War. The car was also the beneficiary of a particularly felicitous moment in automobile history. Under Bruno Sacco's strict leadership, the design stressed not ostentation or trendy novelty, but rather a considered balance between design and engineering, with perhaps a slight tilt in favor of the latter.* In my mind, that made it a classic. The 300E was definitely popular—more than 2.7 million were produced over thirteen years, and it was Daimler-Benz's best-selling model up to that time.

WE HAVE BECOME USED to cars that never fail to start, never falter, never break down. It was not always so. I remember my

* Sacco retired in 1999, but already by the mid-1990s, Mercedes management was relaxing his disciplined approach and adopting a looser design philosophy, not always to good effect. To my eyes, the present trendy generation of the Mercedes G-class, for example, is hardly distinguishable from other high-end SUVs.

old BMW 1600, whose distributor I often had to open and wipe dry in damp weather, and whose engine—not fuel-injected—was often balky in the cold Canadian winters. I always carried jumper cables and spare fan belts in the trunk, and I had a battery charger at home. Over the years, the reliability of automobiles has improved greatly; occasional oil changes are all that is required. Once in a long while, you get a flat tire, but the days when drivers carried a patching kit for repairing punctures are long gone; many cars today don't even have a spare tire. But reliability doesn't necessarily mean durability. "The 1995 Camry was probably the best car in the world when it was new, and I never knew anybody to have serious trouble with them," wrote Jack Baruth in *Road & Track*. "But one by one they've simply disappeared. They were engineered very thoroughly to meet certain durability targets. Having met them, they fade away into the junkyard like forgotten soldiers." Nor does durability always translate into reliability. Baruth explicitly mentions the Mercedes 300E, and makes the point that some of its mechanical innovations could be unreliable, its complicated electricals could create knotty problems, and maintenance was expensive. All true. I've often replaced windshield wiper blades on cars, but never the whole wiper assembly, as I had to do on our Mercedes, when the ingenious single-wiper mechanism failed—although that was after twenty years. Nevertheless, what was striking about the car was that it ran as smoothly after two decades as on the day we bought it. "Good as new" is a common expression, but it really did apply; the controls and switches always felt solid, the MB-Tex lasted as I had expected, and after a car wash the body paint glowed. We passed 100,000 miles without noticing, then 150,000. Durability. And our car lived up to Bruno Sacco's second law; new Mercedes models didn't make it look old. Quite the opposite; she was an elegant old lady.

But time catches up to us all. Having reached a certain age ourselves, Shirley and I decided it was time to downsize and move from our old stone house in a garden suburb of Philadelphia to something smaller and more manageable downtown. Although our new loft came with a garage space, it seemed likely that we would not need a car, and we discussed selling the car after we relocated. As it turned out, the old girl had other plans, which she revealed the day before the scheduled move: a sudden failure of the left ball joint. Ball joints are part of the front suspension—the hip joints of a car—a complex assembly that not only allows the front wheels to move up and down independently of one another, but also makes both wheels turn left or right together. The twenty-four-year-old Mercedes had been sending me signals for the previous couple of weeks that something was wrong, but, caught up in the psychodrama of our move, I had ignored the warnings. Eventually, unable to continue, the joint failed; thankfully, the faithful machine collapsed at a stop sign rather than on the highway.

We rented a Suzuki from Enterprise for a few days, and after selling the Mercedes, we never replaced it. Shirley pointed out that my reflexes were not what they once were, so it was just as well to stop driving. I agreed, but it was a change. When I was a boy in postwar England, the family traveled on bicycles—my father had his name on a waiting list for a car—but after we moved to Canada, there was always a car in the driveway. I remember a British Ford Zephyr before the Vauxhall Velox—my father was a resolute Anglophile when it came to cars, although late in life he drove a Honda Accord. It was the family holiday motor trips that I remember best, the campsites in state parks and the roadside attractions: Ausable Chasm and Frontier Town, Santa's Village and the Desert of Maine, New Hampshire's Old Man of the Mountain. When I was old enough to drive, I was sometimes allowed to take the Vauxhall out alone, an excuse to speed on backcountry roads. I did not own a car as a college student, but ever since the German VW, I'd always had a car—driving to work, driving to shop, driving to movies at the mall, driving to visit family and friends, driving on holiday trips. Always driving.

Don't you miss not having a car? friends ask me. Not really, I answer. Shops and restaurants are within easy walking distance; so are the wine store, the pharmacy, and the bank—the retirees' circuit. Philadelphia has a convenient bus system, and of course we were not really carless. I remember Shirley saying, years ago, "Wouldn't it be great to have a chauffeur?" Well, now we did: Herr Uber. And she was right. Being driven meant not worrying about changing the oil, fretting about repairs, or remembering state inspections. Equally important, it meant never having to navigate to an obscure destination or looking for a parking spot. Somebody else's job.

CHAPTER
ELEVEN

THE NEXT CAR

Electricity is the thing. There are no whirring and grinding gears with their numerous levers to confuse. There is not that almost terrifying uncertain throb and whirr of the powerful combustion engine.

—THOMAS A. EDISON

IN 2022, CALIFORNIA PASSED A REGULATION BANNING THE sale of new gasoline-powered vehicles by 2035. In short order, five states, including New York, Massachusetts, and New Jersey, followed suit, ten other states announced they would adopt similar legislation, and the Biden White House issued an executive order requiring half of all new vehicles to have zero emissions by 2030.

Given the current state of automobile technology, the next car seems likely to be electric. As we have seen, the electric car has a long history. In 1894, at the same time as Ferdinand Porsche was working on the Egger-Lohner electric car in Vienna, two Philadelphians, Henry G. Morris, a mechanical engineer, and Pedro G. Salom (1831–1912), an electrochemist, patented an electric car. Two years later, they produced the Electrobat, an electrically powered taxicab. Two 1.5-horsepower electric motors powered the front wheels, steering was by the rear wheels, and a forty-eight-cell battery gave a range of up to twenty-five miles. The body resembled a traditional hansom cab, with the driver sitting behind

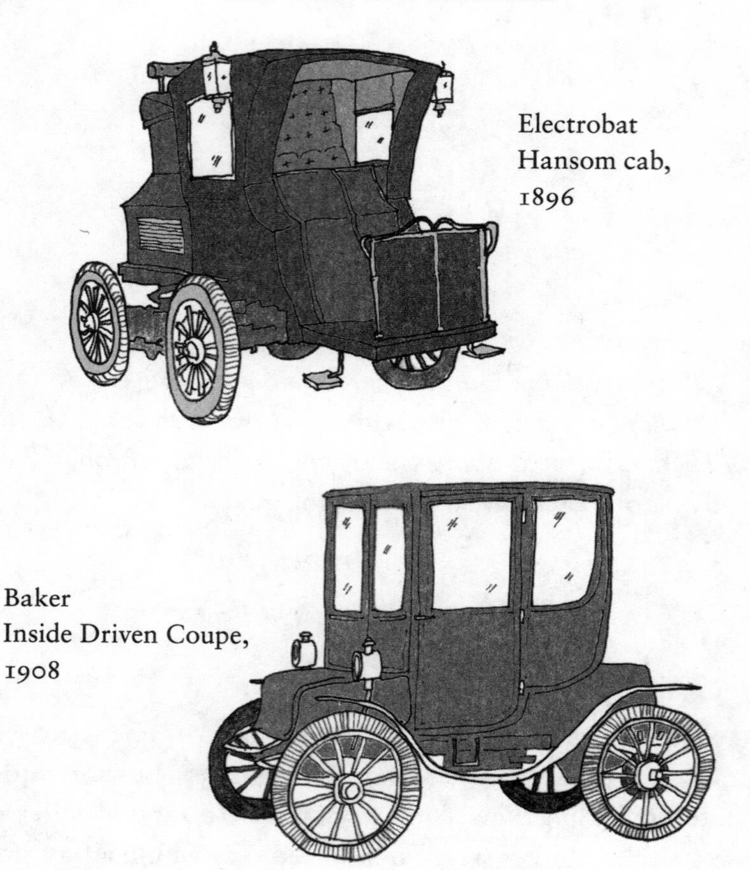

Electrobat Hansom cab, 1896

Baker Inside Driven Coupe, 1908

and above the passengers. The car was controlled by a steering lever and a hand throttle; there was a brake pedal and a pedal that rang a bell to warn pedestrians and horse-drawn carriages. The battery box below the driver was designed to be swapped out with a fresh one at the taxi garage. Morris and Salom sold their invention to the Electric Vehicle Company of New Jersey, which manufactured several hundred cabs and operated them in Boston, New York, Philadelphia, and Washington, DC.

Due to the limited availability of electricity, the first electric vehicles were confined to cities, chiefly to commercial uses. For example, New York Edison and Harrods department store in London had their own fleets of electric delivery vans. The Baker

Motor Vehicle Company of Cleveland, which called itself the "oldest and largest manufacturer of electric motor cars in the world," made an assortment of passenger cars in addition to vans and trucks. Thomas Edison's first electric car was a Baker—he and engineer Walter C. Baker (1868–1955) were friends, and Edison later designed a rechargeable nickel-iron car battery for Baker. Because electric cars did not require laborious hand-cranking, they were considered particularly suitable for women drivers; the Taft White House acquired a Baker electric runabout for Helen Taft's use—and that of three successive First Ladies. The 1908 Baker Inside Driven Coupe was a luxury vehicle for the well-to-do lady driver, selling for $2,800 at a time when a Model T cost well under $1,000. The enclosed cabin—a rarity when most electric cars were open—resembled an elegantly outfitted sitting room, with fabric wall coverings, wool broadcloth upholstery, a makeup mirror, window blinds, and a flower vase. The two facing sofa-like bench seats carried four. Instead of a steering wheel, there was a discreet foldaway tiller. With twelve nickel-iron batteries—six in the front and six in the rear—the Baker had an impressive range of up to one hundred miles, and a top speed of twenty-five miles per hour. The headlamps were electric, as was the interior lighting.

Electric cars faced many challenges. Charging was available in large cities, but electrification was absent in rural areas, precluding long-distance travel. Meanwhile, gasoline service stations were beginning to proliferate, and Ford's Model T drastically reduced the cost of car ownership. But the death knell of the electric car—and the steam-powered car—was hastened by the invention of the electric starter. After a prominent automobile executive was killed when the engine of a car that he was hand-cranking misfired and he was struck by the crank, in 1911 Cadillac commissioned the electrical engineer Charles F. Kettering (1876–1958) to develop a self-starter. Kettering was a prodigious inventor, second only to Edison, and responsible for such significant innovations as Freon refrigerant and leaded gasoline. He had once worked for the

National Cash Register Company, and he adapted the magnetic relay of a cash register, using it not only to provide the spark for ignition but also power for the headlights. Kettering's patented system was immediately adopted by Cadillac, and soon by other car manufacturers.

There was one last attempt to manufacture a mass-market electric car. Edison, who had built an electrically powered automobile as early as 1895, was always on the lookout for new markets for his nickel-iron batteries. In 1914, he convinced Henry Ford, who was basking in the success of the Model T, to build an electric car. Ford's engineers produced several working prototypes using a Model T chassis. "The fact is that Mr. Edison and I have been working for some years on an electric automobile which would be cheap and practicable," Ford told the *New York Times*. "The problem so far has been to build a storage battery of light weight which would operate for long distances without recharging." But batteries, whether nickel-iron or lead-acid, proved both inadequate and too heavy, and even Edison was unable to come up with a workable alternative. Within two years, Ford abandoned the project.

It would be another thirty years before an electrically powered car reappeared, although it was hardly a car at all in the conventional sense. According to most accounts, the golf cart came about thanks to Merle Williams of Long Beach, California. In 1946, during the Second World War when gasoline was rationed, Williams founded the Electric Marketeer Manufacturing Company, which produced electric bicycles that towed one-wheeled trailers "to the market." Later, as a favor to his wife, Peggy, Williams built an electrically powered runabout, and like the British inventor Lawrence Bond, whose design for a shopping vehicle for his wife turned into the Bond Minicar, Williams's electric cart became the foundation of a successful business. The three-wheeled vehicle was steered by the front wheel, but unlike the Bond Minicar, power went to the rear wheels. Williams did not try to turn the cart into a roadworthy minicar, but marketed it as on-site transportation for cargo and people in large industrial plants, airports, resort hotels, and rest homes.

In 1951, Williams introduced a new model that he called the Caddie Car. The basic vehicle consisted of a frame supporting a bench for two golfers, with space in the rear for their bags. There was no body to speak of, no doors or windshield, only a lightweight shade canopy. The front wheel was steered by a tiller, and the two rear wheels were powered by a four-and-a-half-horsepower electric motor giving a maximum speed of about ten miles per hour; lead-acid batteries provided a range of eight to ten miles. Electric golf carts proved an instant hit—they suited short trips over the greens, battery charging was available at the clubhouse, and the quiet vehicles did not disturb other players. When he retired, Williams sold his company to Westinghouse.

Marketeer golf cart, 1960

Golf carts turned out to have a wider application than only golf courses. In 1959, Joel H. Cowan (b. 1936), a young Georgia developer, founded the planned community of Peachtree City. The plan included not only a golf course, but an extensive network of dedicated cart paths, which transformed the golf cart into a sort of neighborhood minicar. Useful for errands and local trips, golf carts were also popular with students—the local high school parking lot eventually provided a special section just for carts. Peachtree City was the model for what became known as "golf cart communities"—not merely places with golf courses but places where the easy-to-operate and inexpensive golf cart was the prime mode of local transportation.

For the next several decades, *electric car* basically meant "golf cart."* That changed in 1996 as a result of a California environmental law that required automobile manufacturers to sell a certain number of zero-emission vehicles by 2001. In response, General Motors test-marketed an electric car, the EV1, in Los Angeles, Phoenix, and Tucson—not selling, but leasing the cars. The EV1 was a subcompact two-seater with a front motor and front drive, an aerodynamic body, and lead-acid batteries providing a range of fifty-five miles (nickel-metal hydride batteries in later models doubled that range). Honda and Toyota had similar programs, but these experiments were terminated, in part because of the cars' limited range and slow charging capacity, in part because gas prices remained relatively low, and mainly because, after intensive lobbying by car makers, California dropped its mandate.

Toyota's electric car was a modified version of its popular compact sport-utility vehicle, the RAV4, but the company was concurrently working on a different type of low-emission car. Almost a century earlier, Ferdinand Porsche's Semper Vivus had demonstrated that one way to overcome the limited range of an electric car was to provide a backup internal combustion engine. The Toyota Prius, introduced in Japan in 1997 (and three years later in the United States), was the first mass-market hybrid. The Semper Vivus was a "series hybrid"; that is, the gasoline engine did not power the wheels, but charged the battery, as compared to a "parallel hybrid," in which both the electric motor and the gasoline engine, separately or together, power the car. The Prius combined both systems: its fifty-seven-horsepower gasoline engine could be used to independently charge the batteries while on the go, and

* It also meant "moon buggy," the electrically powered Lunar Roving Vehicle built by Boeing for NASA's 1971–72 Apollo program. Three LRVs are still parked on the moon.

the engine could also power the front wheels, working independently or together with a forty-horsepower electric motor. At the same time, regenerative braking recovered the kinetic energy of braking, normally lost as heat, and stored it in the battery. (In 2012, Toyota produced a plug-in version of the Prius in which the battery could be topped up from an external source such as a household outlet or a charging station.) The Prius was a four-door compact, although because of the nickel-metal hydride batteries, the car weighed considerably more than a typical compact. This, and the relatively small engine, limited acceleration and top speed. Fuel consumption was about fifty miles per gallon.

Toyota Prius, 2004

The first-generation Prius resembled a generic Japanese sedan. The second generation, which arrived in 2004, had a redesigned body, a compact liftback with a streamlined shape, and a very low drag coefficient of 0.26. *Car and Driver*, which named the 2004 Prius its Car of the Year, called it a "wheeled lozenge." While not exactly beautiful, the distinctive body effectively identified its owner as a green advocate. The truncated rear of the Prius was a Kammback, named after Wunibald Kamm (1893–1966), a pioneering German automotive engineer and aerodynamicist who, in 1938, discovered that tapering and abruptly truncating the rear of a car reduced turbulence more effectively than the customary teardrop shape. Since first being used in a 1940 BMW roadster, the Kammback has appeared both in high-performance two-seaters and family cars from Audi, Citroën, and Honda.

Nissan Leaf, 2010

The first fully electric mass-market car appeared in 2010. The compact Nissan Leaf was not a lozenge but a familiar hatchback. The 110-horsepower electric motor of the Leaf, which had the same wheelbase as the Prius, was located in the front and drove the front wheels; the six-hundred-pound battery pack was under the seats. Like the Prius, the Leaf had regenerative brakes, and like all electric cars, it had no transmission—there were no gears; power went directly from the motor to the wheels.

"Fear extends to and permeates the ownership experience," concluded *Car and Driver* after a three-month test of the Leaf. The magazine determined that the average range of the car was a measly seventy miles. "You're afraid you won't make it to the next electrical outlet, afraid of having to take a charge-sapping detour to buy milk, afraid to turn on accessories like the climate control or the radio." For the manufacturer, an electric car was a delicate balancing act: greater range required more batteries, which meant even more weight, which meant a larger motor, increased cost, and so on. On top of its limited range, the Leaf cost $10,000 more than a Prius.

THE LEAF AND THE PRIUS were answers, or partial answers, to the problem of emissions, but they were not particularly appealing cars, either in terms of performance or design. That changed thanks to the Tesla Roadster. Tesla Motors was a Silicon Valley startup established in 2003 by Martin Eberhard (b. 1960), an electrical engineer, and Marc Tarpenning (b. 1964), a computer scientist. Although neither had automotive experience, they set out to manufacture an electric car. Their largest investor was the South African–born billionaire entrepreneur Elon Musk (b. 1971), who had made his fortune with PayPal, an online payment system, and had recently founded SpaceX, a spacecraft manufacturer. Musk took an active interest in Tesla, and eventually replaced Eberhard as head of the company.

Tesla's strategy was to make a car that would alter the public's perception of electric cars as glorified golf carts. "With the Roadster, we wanted to create a product that would break the mold and convince people that an electric car could be cool, sexy, and fast," said Musk. The limited-production, high-performance Tesla Roadster was a two-seater developed in collaboration with the well-regarded British sports car maker and engineering consultancy Lotus Cars. The Roadster was built on the modified chassis of the Lotus Elise, a two-seater introduced in 1996. The carbon-fiber body of the Tesla, which in broad strokes resembled the Elise, was largely the work of the Lotus design team, headed by Donato Coco (b. 1956), an Italian who had worked for Citroën and Ferrari. The body, on an extruded bonded-aluminum chassis, was built in Britain and shipped to California, where the power train was installed. One of the key innovations of the latter was the battery pack, which consisted of 6,831 off-the-shelf lithium-ion cells, the tiny rechargeable batteries normally used in laptop computers. Lithium-ion batteries, not previously used in a production car, were lighter and more efficient than traditional nickel-metal hydride batteries and gave the Roadster a range in excess of 240 miles, a first for a production electric car. The Tesla equaled—and in some respects surpassed—the Lotus Elise in terms of performance, thanks to the transmission-less electric motor that delivered power with unparalleled immediacy, making for exceptional acceleration (zero to sixty miles per hour in 3.7 seconds) and a top speed of 125 miles per hour. The car cost $98,000, but Tesla announced plans to build a more affordable sedan.

Tesla Roadster, 2008

Only 2,500 Roadsters were built, but that was enough to demonstrate that an electric vehicle could be an exciting and desirable automobile, although it was still not clear to many whether Musk was capable of delivering a mass-market electric car or was a latter-day Preston Tucker. In 2010, Tesla Motors received a $465 million loan from the federal Department of Energy and acquired a disused Toyota assembly plant in Fremont, California. That same year, the company went public. Two years later, the Tesla Model S rolled off the assembly line.

The Model S, which sat five, was a full-size "four-door coupe," a luxury category that emphasized performance and style. The iconoclastic Tesla offered practicality as well, with an optional third row of rear-facing jump seats for two children, similar to a 1960s station wagon. A very fast wagon: the 416-horsepower rear-mounted electric motor driving the rear wheels pushed the car to a top speed of 134 miles per hour with zero-to-sixty acceleration almost as fast as a Porsche 911. The massive floor-mounted battery, which weighed 1,300 pounds, had a range in excess of two hundred miles, depending on driving speed and outside temperature.* The Model S established Tesla as a leader in the electric car field. In 2013, *Motor Trend* magazine named it Car of the Year, and in 2015 and 2016, despite its hefty price, the Model S was the world's best-selling battery-electric car, ahead of the anodyne Leaf.

* Cold weather can reduce the range of an electric vehicle by as much as thirty percent. The cold also increases charging times. Teslas heat and cool the battery pack to reduce these effects.

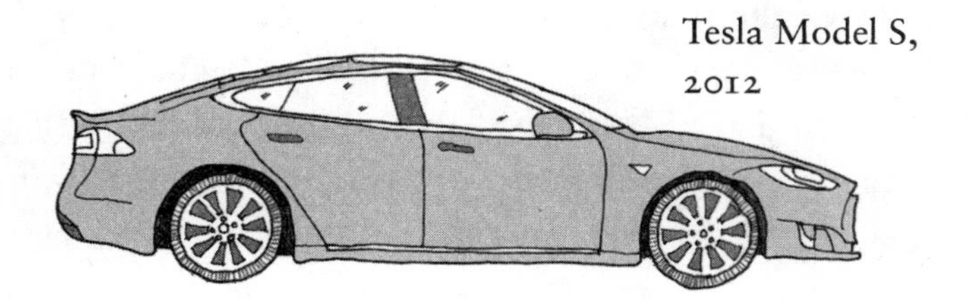

Tesla Model S, 2012

The Tesla Model S was a revolutionary car, not only because it was electric but also because it effectively functioned as a smartphone on wheels. This was the first automobile whose performance was defined by its software—a car that could communicate with its owner, and more important, a car that could communicate with its maker. Like a smartphone, the Tesla's computer received regular upgrades that improved handling, ride, performance, and safety. This feature was as industry-changing as Ford's moving assembly line and Sloan's strategy of dynamic obsolescence.

Maintaining up-to-date functionality not only improved the ownership experience, it also affected the car's resale value. "This car actually gets better with age," said the Model S's designer, Franz von Holzhausen (b. 1968), who had joined Tesla in 2008. Born in Connecticut, he was a graduate of the ArtCenter College of Design in Pasadena, whose pioneering transportation program, founded in 1948, was responsible for educating many leading car and motorcycle stylists. Holzhausen had experience working for European, American, and Asian manufacturers. After spending eight years in Volkswagen's California studio, he moved to General Motors' new design studio in North Hollywood, and later became chief of design for Mazda of North America. He could be critical of the legacy car makers. "There's a pretty consistent problem of over-managing. What we don't have at Tesla is layers and layers of decision making," he told an interviewer. "And we're not caught up in the brand . . . if it's the right thing to do, it's the right thing."

The right thing at Tesla was a blend of function and form.

The Tesla aesthetic owed a lot to the sporty European cars of the 1960s, with their simple shapes that respected but were not driven by wind tunnel testing, and whose distilled designs had little chrome trim and extraneous detail. The minimalism of the Tesla S stood in sharp contrast to the prevalent car design ethos, which favored exaggerated sci-fi forms that seemed derived from Transformers movies.

In 2015, Tesla unveiled the Model X, a luxury crossover. As has been mentioned, the traditional sedan had been largely replaced by the sport-utility vehicle, especially in the United States. Some SUVs were referred to as crossovers, because they were more like cars than trucks—effectively, tall hatchbacks with large interiors. The Model X, which was built on the Model S platform, had several unusual design features: a windshield that curved seamlessly over the driver's head, and a third row of forward-facing seats that turned the car into a speedy seven-passenger minivan. Instead of depending on traditional door handles, the motorized doors could be opened and closed remotely, as the driver approached the vehicle. The rear doors were the Model X's most unusual feature: large rear gull wings. These motorized double-hinged doors, fitted with sensors to avoid hitting an adjacent car or a low ceiling, were marvels of technology, although they reminded me of the gull-wing doors of the short-lived 1981 DeLorean and seemed like a case of design overreach.

The promised "affordable Tesla" finally arrived in 2017. While the base price of a Model S was $71,000, and the Model X cost $80,000, the Model 3 could be had for $36,000. With the government rebate on electric vehicles available at that time, the price was below $30,000, which compared favorably to midsize hybrids from Toyota and Honda. Smaller than the Model S, weighing almost a thousand pounds less, with a 346-horsepower motor and a smaller battery pack, the Model 3, which sat five—snugly—had an EPA-estimated range of 220 miles (a version called the Long Range increased this to 310 miles).

Like the Model S, the Model 3 was a four-door fastback sedan,

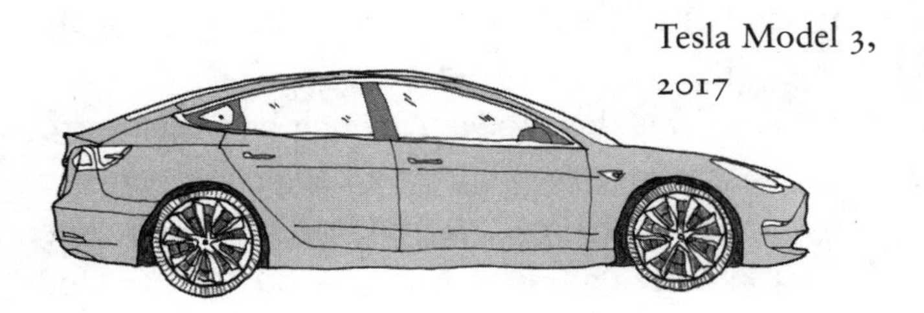

Tesla Model 3, 2017

and it shared some of the former's design features, such as a tinted glass roof (though not openable) and a streamlined body with a low drag coefficient. With a rear engine and no conventional radiator, there was no need for a grille, and unlike the first Model S, which had a vestigial grille, the front of the Model 3 was a blank concavity that resembled a toothless smile. The Kammback rear was similar, too, although a trunk lid replaced the liftback; the flush door handles were simplified versions of those in the Model S. The interior was an exercise in Bauhaus minimalism, with no visible knobs or switches. Virtually all the controls were displayed on a thin fifteen-inch touch screen mounted on the blank dash. Whether you found the interior austerely Zen-like, or simply bare, was a matter of taste. The aesthetic reminded me of a Braun slide projector I owned in the 1960s: a plain box with a circular opening for the plastic screw-in lens, a slit for the manually operated slide holder, and a removable ventilation grille so that you could change the projector bulb.* That was it. What you saw was all there was.

"SO, HOW DO YOU like your car?" I asked my friend Jason—we were driving in his new Tesla. By way of an answer, he floored it. *Oof!* I had read of the sensation of being pushed back into your seat in racing cars—or fighter jets—but I had never actually experienced it. Until now. We were on the Benjamin Franklin Parkway

* The Braun D10 slide projector was designed by Dieter Rams in 1962.

in downtown Philadelphia, so he had to almost immediately slow down for a light, but I got the idea.

Jason's Tesla was the Model Y, which had been introduced in 2020, a compact crossover built on a Model 3 platform and sharing many parts with that car, but slightly taller and wider. Just as the Model 3 had been a less expensive Model S, the Model Y was a less expensive Model X—the price in 2020 was $51,190. The design was simplified—no windshield extending into the roof and no complicated gull-wing doors. In the first quarter of 2023, shortly after Jason bought his car, the Model Y became the world's best-selling car—electric or gasoline-powered—ahead even of the ever-popular Toyota Corolla.

Tesla Model Y, 2020

"I'm not really a sedan guy. I like station wagons," Jason told me. "Maybe because I grew up with my father's Volvo." Jason lives in downtown Philadelphia. He and his wife both walk to work. "We're a one-car family. Angie rarely drives, and both girls are away at college. I'm the one who uses the car." Jason's real estate development company has projects outside the city, and he often transports presentation boards and easels to town meetings. His last three cars were station wagons: a Volkswagen Passat, a Subaru Outback, and an Audi A6 Allroad. When the Audi reached 100,000 miles, he decided it was time to switch to an electric car. There was only one electric station wagon on the market: the expensive Porsche Taycan. Volvo's plug-in hybrid wagon was pricey, too. Audi made an electric SUV, but Jason didn't fancy an SUV. The Tesla Model Y crossover seemed like it might be a good compromise—not quite a wagon, but not really an SUV, either. "I figured that Tesla would be more knowledgeable about EVs than the established car makers," he told me.

Jason visited a Tesla store, which was in the King of Prussia Mall. The Model Y was out with another customer, so he test-drove a Model 3. The salesman handed him the key—actually, a plastic card—and told him, "Go have fun. Bring it back in half an hour." Jason drove through nearby Valley Forge Park, where he tried out the autopilot. "The technology really wows you," he told me. The minimalism of the interior appealed to him, and he appreciated the lower environmental impact of an electric car. He didn't actually get to drive the Model Y, but the salesman assured him it was similar in most respects. "I thought the Model 3 was nicer-looking, but the Model Y was more spacious and it had a large liftgate," he said. The Model Y came with all-wheel drive, like his Audi, which also appealed to him because the family went skiing in the winter. Jason ordered the car online. There were five color options. "I went with white, mainly because it had the earliest delivery date." Nine months later, in mid-December, he got an email that his car was ready to be picked up at Tesla's suburban facility.

"My first impression was that the Model Y wasn't quite as well finished as my Audi; it didn't have the same quality of detailing and the feeling of rock-solid durability," he recalled. What about the controls, I asked? "I got used to the touch screen pretty quickly. I guess that buyers of the Tesla skew younger, and we've grown up with smartphones and iPads. Although I do wish that the screen wasn't so fussy—it can be a little distracting," he said. "The regenerative braking took a little longer to master. The instant you take your foot off the accelerator, the car starts braking. It makes for a pretty jerky ride until you get used to it. The best thing to do is to keep the pedal slightly depressed, or use the autopilot. I didn't get the enhanced autopilot. The standard autopilot keeps the car in the lane, changes lanes, and maintains speed and distance from other cars. It's pretty useful for highway driving, and it's good in stop-and-go traffic, too."

To prolong battery life, Jason limited his charge to eighty percent, which typically gave him a range of about 260 miles. "I generally drive no more than sixty to a hundred miles a day, which is well within that range. I installed a 240-volt charger in our garage that gives me about thirty miles for each hour of charging, so overnight produces a full charge." Jason's first long car trip was to Scranton, about 125 miles from Philadelphia, or 250 miles for the return trip. "That was getting close to my limit, so I looked up charging stations in Scranton on my phone app and found that there were two in the Hilton where I was staying. When I got there, the hotel charging cable had a non-Tesla plug. I remembered something about my car coming with adapters, and I found one in the trunk. The next morning, my car was fully charged."

Jason had few complaints. "I've had the car for seven months and there haven't been any major problems." The computer occasionally had glitches—the autopilot could turn off unexpectedly—and the phone's voice-command function could sometimes be erratic. Jason didn't appear to be put off by these lapses. Unlike some of my friends, who seem to fall in love with their Teslas, Jason was pretty levelheaded. "My Audi was a great car with a lousy computer. The Tesla is an okay car with a great computer."

Was there anything special that Jason liked? "I know you're not supposed to do this, but when I'm stopped at a red light, I often check my phone messages," Jason confessed. "If you get immersed in doing this, you sometimes get startled by the honking of the impatient driver behind you. My Tesla recognizes when the light turns green and sounds a ding to remind me. It's a little thing, but I really appreciate it."

A little thing. I understood Jason's comment; whenever I drove the Mercedes, I always enjoyed the view of the hood ornament, which was like a little gunsight, aiming the car down the open road. Driving an automobile, whether it's powered by an internal combustion engine or an electric motor, is an intimate experience. Driver and machine. Driving provides an opportunity for the exercise of personal control, as well as a feeling of movement and speed, and above all a sense of freedom. This was equally true of the automobile's predecessor. "A swift carriage, of a dark night, rattling with four horses over roads that one can't see—that's my idea of happiness," Isabel Archer, the willful young heroine of Henry James's *The Portrait of a Lady*, confesses to a friend. She means it as a metaphor, but it is telling nonetheless. In James's time, there was an assortment of personal carriages—gigs and traps, phaetons and surreys, barouches and landaus—fast and slow, lavish and functional, sporty and workaday. Yet the end of the horse-drawn era was coming, foretold on that fateful summer's day in August 1888 when Bertha Benz took her husband's noisy contraption on her famous sixty-five-mile ride. I imagine the indomitable Bertha adjusting the carburetor, grasping the steering handle with one hand and the power lever with the other, and calling out to Eugen and Richard, "Push, boys, push!"

CODA

ELEGY

I'm in favor of progress; it's change I don't like.

—MARK TWAIN

IN 2022, *CAR AND DRIVER* SURVEYED TWENTY-TWO HEADS of design at leading American, European, and Asian automobile companies and asked them to name the ten most beautiful cars of all time. Of the more than two hundred automobiles selected, only ten had four or more votes. All were sports cars; one dated from the 1970s, one from the 1950s, and two from the 1930s. The rest were all from the 1960s, including the top four: the Porsche 911, the Corvette Stingray, the Lamborghini Miura, and the number one pick, the Jaguar XKE, which Enzo Ferrari is reputed to have called "the most beautiful car ever made."*

The sixties were a special decade in car design. On the one hand, most of the technologies that defined performance were available. For example, the Jaguar had partial synchromesh transmission, four-wheel disc brakes, rack-and-pinion steering, and independent front and rear suspension. Yet the decade was early enough in the history of the automobile that there was still a sense of excitement, of unexplored frontiers. The 1960s—and the late 1950s—were a particularly inventive period that produced not only beautiful sports cars but sedans such as the Mercedes-Benz 180, BMW's Neue Klasse, and the Citroën DS; off-road vehicles

* The other two 1960s cars were both Ferraris: the 206 GT and the 250 GTO.

such as the Jeep Wagoneer and the Toyota 40 series Land Cruiser; as well as small cars such as the Mini, the Fiat Cinquecento, the Renault 4, and the Corvair.

It is worth noting what was *not* present in the sixties. The global market, which would eventually homogenize car design, had not yet developed. Neither had the computer-aided design tools that today allow designers to create the complicated folds and scoops that result in cars that display a comic book idea of beauty. The fact that many 1960s car designers had engineering backgrounds, rather than design school training, was probably a factor. Or perhaps it's just that today's buyers have a rigid idea of what they want—SUVs, hatchbacks, or minivans—leaving the designer little leeway except to fiddle with superficial creases and trim.

My first car, the VW bug, was conceived in 1938, and my last in 1984. The intervening five decades saw the introduction of most of the features that defined the modern automobile: flashing turn signals, disc brakes, radial tires, and automatic transmissions in the 1940s; monocoque construction, air-conditioning, power steering, tubeless tires, and three-point seat belts in the 1950s; electric windows, intermittent wipers, alternators, dual-action tailgates, and head restraints in the 1960s; catalytic converters, airbags, and antilock brakes in the 1970s; and fuel injection in the 1980s.

The driving machines of that period were mechanical. Speedometers, tachometers, and gauges were analog; controls were stalks, knobs, and switches. The engine was managed by carburetors and simple fuel injection. Gear shifting, braking, and steering were achieved by hydraulic and mechanical linkages, both unassisted and powered. Analog cars not only provided the driver with a tangible sense of control, but were relatively easy to understand, and hence to maintain and repair. Car controls were uncomplicated and universal: a keyed ignition (introduced in the 1940s), three pedals, a gearshift, a hand brake, and one or two stalks for turn signals, headlights, and windshield wipers. It was possible to get into an unfamiliar car—whether it was a rented

Renault 4 or a girlfriend's father's Thunderbird—and immediately feel at home. There was no learning curve.

The twentieth century was not only the American Century, as Henry Luce called it, it was also, as the historian John Lukacs pointed out, the Automobile Century. Personal mobility, the car's chief virtue, altered where and how we lived. As a schoolboy, I rode in Montreal streetcars—they were cream-colored with red trim and rattan seats. Streetcars and railroads produced linear urban development, but cars could go in all directions. Thus, old cities sprawled into the surrounding countryside and new cities took unprecedented low-density forms. Individual mobility also created new destinations: drive-in movie theaters and drive-through banks, shopping malls and big box stores, and far-off weekend cottages.

There was a price to pay for this mobility. The new suburban malls hurt the old downtowns, and spreading out instead of concentrating hurt the old cities, most of which lost population. Life in the new suburban communities was inconvenient for the carless, the very young and very old. The predominance of personal transit—which is what the private car provides—meant the inevitable decline of mass transit such as streetcars and interurban railways. Despite various efforts—better combustion, unleaded gas, catalytic converters—gas emissions contributed to global warming. There were other less evident effects, and not only on the environment. When you stepped off a streetcar or a bus, it continued on its way, but when you reached your destination in a car, you needed a place to park. Thus, parking lots and parking garages became unsightly features of the suburban and urban landscapes.

It is estimated that there are now more than 1.5 billion cars worldwide—one-third of them in Asia—and yet the primacy of the automobile appears to be diminishing in North America. Not that we have given up on the car; far from it. According to the 2020 US Census, 91.5 percent of American households own at least one vehicle, which is up slightly from 90.9 percent in 2015.

Many people have more than one car, and the total number of registered cars went up during that same period. But something has changed. The pleasurable Sunday drive has become a distant memory, like dressing for dinner, and motoring has become more like mowing the lawn or weekly grocery shopping—it's a chore. No one likes commuting, with its traffic jams and frustrating delays, as is evident by the reluctance of many workers to return to the office after the pandemic. Our romantic affair with the automobile has cooled and settled into something that seems more like a loveless marriage.

Analog cars were a key part of that springtime romance, but today they are becoming the automotive equivalent of vinyl records, mechanical watches, and fountain pens: anachronistic throwbacks. Digital controls have replaced mechanical linkages, digital engine management has replaced carburetors, and computers are in charge of steering, suspension, braking, and much more. Instead of analog gauges and dials, touch screens control lights, infotainment, navigation, and heating and cooling. Doors open and close automatically. Cars park themselves and warn the driver of hazards. Multiple cameras peer warily in all directions. If the self-driving car ever arrives, it will be the final capitulation. Whether we end up with hybrids, or cars powered by electric batteries, hydrogen cells, or synthetic fuels, it is evident that a new era is beginning. No more rumbling engines, no more downshifting in corners, no more heel-and-toeing, no more tinkering and tuning. For someone who grew up with the analog car, the digital car is an alien creature: all-knowing, too smart for its own good, a bossy nanny. But that may just be whistling in the wind. As E. B. White wrote of his old Model T, "I suppose it's time to say goodbye. Farewell, my lovely!"

ACKNOWLEDGMENTS

ALTHOUGH I HAVE WRITTEN DESIGN HISTORIES BEFORE—about buildings, chairs, and hand tools—this is not a book I expected to write. I must first acknowledge my late wife, Shirley Hallam, who put up with my rather-too-frequent car purchases, and who over the decades shared many memorable car trips—beginning with an escape from the Montreal winter to Key West in a BMW 1600. This is my first book that does not benefit from her keen editor's eye. My thanks to Jason Duckworth, who generously shared his experiences buying and driving a Tesla Model Y; thanks also to David Burns, another friend who is a Tesla devotee. The mechanical mysteries of the driving machine have been explained to me over the years by many patient mechanics, including Patrick Hébert of Garage Hébert in Hemmingford, Quebec; Phil D'Amico of Roanoke Auto Service in Chestnut Hill, Philadelphia; and Michael Casale of Silver Star Auto Haus in Skippack, Pennsylvania. This is also a good place to acknowledge my fellow travelers of those two early road trips: my Renault 4 companion, Ralph Bergman; and Ed Satterthwaite, with whom I drove that old Volkswagen from Paris to the Valencia Fallas. Matt Weiland, my editor at Norton, was supportive and stimulating by turns. Finally, I must thank my estimable agent, Andrew Wylie, who understood the promise of this somewhat quirky book and provided his expertise at a critical juncture.

APPENDIX

A CAR DIARY, 1964–2017

’64 **Renault 4L**. A rental, but it was my first road trip, so this car belongs here. (1964)

’60 **Volkswagen Beetle.** Hamburg to Valencia via Paris. (1967)

’62 **Mini Cooper.** This retired rally car was a rash purchase. (1969)

’69 **Citroën 2CV.** Four wheels under an umbrella. *C’était magnifique!* Death by winter rust. (1969–72)

’69 **BMW 1600.** In many ways my favorite. (1972–82)

’68 **Volvo 145 station wagon.** A classic. We were building a house and needed a wagon. (1976–77)

’82 **Subaru GL hatchback.** Our first Japanese car. (1982–83)

’83 **Subaru 4WD GL station wagon.** A trusty tool, especially in the Canadian winter. (1983–86)

’76 **Toyota Celica GT.** A brief fling. (1981–83)

’84 **Innocenti Mini 90L.** Shirley’s favorite; chic, and so much fun to drive. (1984–85)

’85 **GMC S-15 Jimmy.** Keep on truckin’. (1985–89)

’83 **Honda Prelude.** This is as close as I got to a sports car. (1987–90)

’89 **Subaru 4WD GL station wagon.** A second dalliance. (1989–95)

’86 **Audi 4000.** German quality again . . . and heated seats. (1990–93)

’95 **Infiniti G20.** Almost as nice to drive as the old BMW, but a tight fit. (1995–96)

’93 **Mercedes-Benz W124 300E 2.8.** When a Mercedes was a Mercedes. (1996–2017)

FURTHER READING

General

Maurice Alley, "European Postwar Cars," *SAE Transactions* 61 (1953), 503–28.

Edson Armi, *The Art of American Car Design* (University Park: Penn State University Press, 1988).

Stephen Bayley, *Sex, Drink and Fast Cars* (New York: Pantheon, 1986).

Markus Caspers, *Designing Motion: Automotive Design 1890–1990* (Basel: Birkhäuser, 2017).

David Gartman, *Auto-Opium: A Social History of American Automobile Design* (New York: Routledge, 1994).

David Halberstam, *The Reckoning* (New York: William Morrow, 1986).

Russell Hayes, *The Big Book of Tiny Cars: A Century of Diminutive Automotive Oddities* (Beverly, MA: Motorbooks International, 2021).

Thomas Hine, *Populuxe* (New York: Knopf, 1996).

John Lukacs, *Outgrowing Democracy: A History of the United States in the Twentieth Century* (Garden City, NY: Doubleday, 1984).

Ralph Nader, *Unsafe at Any Speed: The Designed-In Dangers of the American Automobile* (New York: Grossman, 1965).

L. J. K. Setright, *Drive On! A Social History of the Motor Car* (London: Granta, 2002).

Penny Sparke, *A Century of Car Design* (Hauppauge, NY: Barron's, 2002).

Gary Witzenburg, "The 10 Most Beautiful Cars according to Leading Automotive Designers," *Car and Driver*, updated November 21, 2022.

Car Makers

Stephen Bayley, *Harley Earl and the Dream Machine* (New York: Knopf, 1983).

Jerry Burton, *Zora Arkus-Duntov: The Legend Behind Corvette* (Cambridge, MA: Bentley, 2002).

Vincent Curcio, *Chrysler: The Life and Times of an Automotive Genius* (New York: Oxford University Press, 2000).

Harley J. Earl, "The Secret of Making Beautiful Cars in the 1950s," *Saturday Evening Post*, August 7, 1954.

Dante Giacosa, *Forty Years of Design at Fiat* (Milan: Automobilia, 1979).

Nik Greene, *Bruno Sacco: Leading Mercedes-Benz Design 1975–1999* (Ramsbury, UK: Crowood Press, 2020).

Peter Grist, *Virgil Exner: Visioneer: The Official Biography of Virgil M. Exner, Designer Extraordinaire* (Poundbury, UK: Veloce, 2021).

Lee Iacocca and William Novak, *Iacocca: An Autobiography* (New York: Bantam, 1984).

Robert Lacey, *Ford: The Men and the Machine* (Boston: Little, Brown, 1986).

Ivan Margolius and John G. Henry, *Tatra: The Legacy of Hans Ledwinka* (Harrow, UK: SAF, 1990).

Masaaki Sato, *The Honda Myth: The Genius and His Wake* (New York: Vertical, 2006).

Motor Trend Channel, "Tesla Chief Designer Franz von Holzhausen," *The InEVitable*, season 7, episode 1, YouTube video, 1:59:02, March 30, 2023.

Alfred P. Sloan Jr. *My Years with General Motors* (Garden City, NY: Doubleday, 1963).

Jonathan Wood, *Alec Issigonis: The Man Who Made the Mini* (Derby, UK: Breedon, 2005).

Cars

Jim Allen, *Jeep* (Osceola, WI: Motorbooks International, 2001).

Malcom Bobbitt, *The Volkswagen Bus Book* (Poundbury, UK: Veloce, 2016).

Thomas E. Bonsall, *Disaster in Dearborn: The Story of the Edsel* (Stanford, CA: Stanford University Press, 2002).

Csaba Csere, "Road Test: Mercedes-Benz 300E," *Car and Driver*, May 1986, 47–57.

Philip S. Egan, *Design and Destiny: The Making of the Tucker Automobile* (Orange, CA: On the Mark, 1989).

Mark Howell, "The Chrysler Six: America's First Modern Automobile," *Antique Automobile*, March–April 1972, 26–30.

Byron Olsen, *Station Wagons* (Osceola, WI: Motorbooks International, 2000).

John Reynolds, *The Classic Citroëns, 1935–1975* (Jefferson, NC: McFarland, 2012).

Bernhard Rieger, *The People's Car: A Global History of the Volkswagen Beetle* (Cambridge, MA: Harvard University Press, 2013).

Don Sherman, "1886 Benz Patent Motorwagen Sparked a Revolution," *Car and Driver*, December 1986.

Dan Strohl, "Ford, Edison and the Cheap EV That Almost Was," *Wired*, June 18, 2010.

E. B. White, "Farewell My Lovely!" *The New Yorker*, May 16, 1936.

INDEX

Page numbers in *italics* indicate illustrations.